The Nature of
the Physical Universe

Contributors

MURRAY GELL-MANN

FRED HOYLE

STANLEY L. JAKI

HILARY WHITEHALL PUTNAM

STEVEN WEINBERG

VICTOR F. WEISSKOPF

THE NATURE OF
THE PHYSICAL UNIVERSE

1976 NOBEL CONFERENCE

organized by

Gustavus Adolphus College
St. Peter, Minnesota

edited by

DOUGLAS HUFF
Gustavus Adolphus College

OMER PREWETT
Gustavus Adolphus College

A WILEY-INTERSCIENCE PUBLICATION

JOHN WILEY & SONS, New York • Chichester • Brisbane • Toronto

Library of Congress Cataloging in Publication Data:

Nobel Conference, 12th, Gustavus Adolphus College, 1976.
　　The nature of the physical universe.

　　"A Wiley-Interscience publication."
　　1.　Particles (Nuclear physics)—Congresses.
2.　Science—Philosophy—Congresses.　3.　Cosmology—
Congresses.　I.　Huff, Douglas, 1944-　　　II.　Prewett,
Omer, 1939–　　　III.　Gustavus Adolphus College,
St. Peter, Minn.　IV.　Title.

QC793.N63　1976　　　　500　　　　78-14788
ISBN 0-471-03190-9

Printed in the United States of America

10 9 8 7 6 5 4 3 2 1

To

Richard Q. Elvee

The Contributors

Murray Gell-Mann, Robert Andrews Millikan Professor of Theoretical Physics, California Institute of Technology, received the 1969 Nobel Prize in physics for his work on the theory of elementary particles. He received his bachelor's degree in physics from Yale University in 1948, and his doctorate from the Massachusetts Institute of Technology in 1951. In 1953 he introduced the quantity called "strangeness." In 1961 he suggested the "eight-fold way" classification scheme, and more recently proposed the "quark" as a subunit of elementary particles.

Professor Gell-Mann is a member of the National Academy of Sciences, Vice-President of the American Academy of Arts and Sciences, and a Fellow of the American Physical Society. He has served as a member of the President's Science Advisory Committee and is now a Citizen Regent of the Smithsonian Institution, and Chairman of the Board of the Aspen Center for Physics.

Sir Fred Hoyle is recently retired from the Plumian Professorship of Astronomy and Experimental Philosophy at Cambridge

University. In 1967 he founded and became the first director of the Institute of Theoretical Astronomy at Cambridge. He was elected a member of the Royal Society in 1957, and served as its Vice-President in 1970. He is also a Foreign Associate of the National Academy of Sciences and of the American Academy of the Royal Astronomical Society.

Sir Fred's books range from theoretical sciences, *The Nature of the Universe* and *Frontiers of Astronomy*, through politics and sociology, *Decade of Decision* and *Man and Materialism*, and science fiction, *The Black Cloud, Ossian's Ride*, and *The Fifth Planet*. He has written a space play for children, *Rockets in Ursa Major*, and also is the author of an opera libretto, *The Alchemy of Love*.

Stanley L. Jaki is Distinguished University Professor at Seton Hall University. Professor Jaki, a Catholic priest and member of the Benedictine Order, received his doctorate in systematic theology from the Instituto Pontificio S. Anselmo in 1950 and his doctorate in physics from Fordham University in 1957. He received the Lecomte du Nouy Prize in 1970 for his book, *Brain, Mind and Computers*. In 1974–1975 and 1975–1976 he was Gifford Lecturer at the University of Edinburgh and in 1977 Fremantle Lecturer at Balliol College, Oxford University.

His principal publications are *The Relevance of Physics, The Paradox of Olbers' Paradox, The Milky Way: An Elusive Road for Science, Science and Creation: From Eternal Cycles to an Oscillating Universe, Planets and Planetarians: A History of Theories of the Origin of Planetary Systems*, and *The Road of Science and the Ways to God* (Gifford Lectures). He has also translated the *Cosmological Letters on the Arrangement of the World-Edifice* by J. H. Lambert, and *The Ash Wednesday Supper* by Giordano Bruno.

Hilary Whitehall Putnam is Walter Beverly Pearson Professor of Modern Mathematics and Mathematical Logic at Harvard University. Professor Putnam received a B.A. degree in 1948 from the University of Pennsylvania and a Ph.D. degree in 1951 from the University of California at Los Angeles. In 1951–1952 he was a Rockefeller Foundation Research Fellow, in 1960–1961 a Guggenheim Foundation Fellow, and in 1968–1969 a National Science Foundation Fellow. In 1976 Professor Putnam delivered the John Locke Lectures at Oxford University

Professor Putnam is currently the President of the Philosophy of Science Association and past President of the Eastern Division of the American Philosophical Association. He is also the current President of the Association of Symbolic Logic and a member of the American Mathematical Society. A two-volume collection of Professor Putnam's philosophical papers was published in 1975 under the titles *Mathematics, Matter and Method (Philosophical Papers, Volume I)*, and *Mind, Language and Reality (Philosophical Papers, Volume II)*. In addition to his many journal publications, Professor Putnam is also the author of *Philosophy of Logic* and *Meaning and the Moral Sciences*.

Steven Weinberg is Higgins Professor of Physics at Harvard University and Senior Scientist at the Smithsonian Astrophysical Observatory. Professor Weinberg studied at Cornell University, the Niels Bohr Institute in Copenhagen, and Princeton University where he received his doctorate in 1957. In 1977 he received the Dannie Heineman Prize for Mathematical Physics. He also received the J. Robert Oppenheimer Memorial Prize in 1973, and the American Institute of Physics-United States Steel Foundation Science Writing Award in 1977.

Professor Weinberg is a member of the National Academy of Sciences, the American Academy of Arts and Sciences, the American Physical Society, the American Astronomical Society, and the American Mediaeval Academy. Publications by Professor Weinberg include numerous scientific articles and two books: *Gravitation and Cosmology—Principles and Applications of the General Theory of Relativity*, and *The First Three Minutes—A Modern View of the Origin of the Universe.* He has served as the Richtmeyer Memorial Lecturer of the American Association of Physics Teachers, the Scott Lecturer at the Cavendish Laboratory, Cambridge University, and the Silliman Lecturer at Yale University.

Victor F. Weisskopf is Institute Professor Emeritus and Professor of Physics Emeritus at the Massachusetts Institute of Technology. Professor Weisskopf was born in Vienna in 1908 and received his doctorate from the University of Göttingen in 1931. He is a fellow of the American Physical Society and served as a Director-General of the European Center for Nuclear Research (CERN) in Geneva from 1960 to 1965. In January 1976, he was elected the 35th President of the American Academy of Arts and Sciences.

Professor Weisskopf received the Max Planck Medal of the German Physical Society in 1956 and the Cino del Duca Award for humanism in science. In December 1975 he was named by Pope Paul VI to the 70-member Pontifical Academy of Sciences. He is the author of numerous articles on nuclear physics, quantum theory, and radiation theory, some of which are collected in *Physics in the XX Century*, and he is co-author of *Theoretical Nuclear Physics.* His book, *Knowledge and Wonder: The Natural World as Man Knows It*, was selected by the Thomas Elva Edison Foundation as the best science book of the year for youth in 1962.

Preface

The papers included in this volume were originally presented during the 1976 Nobel Conference at Gustavus Adolphus College, St. Peter, Minnesota. The association between the Nobel Foundation and Gustavus Adolphus College began in 1963 when 26 Nobel laureates gathered to dedicate the college's Nobel Hall of Science as the first American memorial to Alfred Nobel. Two years later the Nobel Foundation gave its endorsement to organize a conference in which scientists could meet to discuss leading topics in science for the benefit of an audience of intelligent laymen.

This year's topic, "The Nature of the Physical Universe," was suggested by Reverend Richard Elvee, who played an instrumental role in bringing both philosophers and physicists together for the purpose of discussing our present understanding of the nature of the physical universe. The papers fall naturally into three general categories: (1) the history and current status of elementary particle theory, (2) the origin and limits of the universe's material resources, and (3) the nature and limits of the scientific enterprise. They can be profitably read by a wide range of people interested or involved in science, but they will be of particular interest to students of philosophy, physics, and the history and philosophy of science.

The first three papers "What is an Elementary Particle?" by Dr. Victor Weisskopf, "What are the Building Blocks of Matter?" by Dr. Murray Gell-Mann, and "Is Nature Simple?" by Dr. Steven Weinberg are an introduction to some of the fundamental ideas underlying our present conception of the elementary particle structure of matter. Professor Weisskopf sketches the development of the conception of the elementary particle from the impenetrably hard atoms of Newton to the conditionally elementary particle of today. Professor Gell-Mann discusses quarks and electron-like particles, two classes of elementary particles many physicists of today believe may ultimately prove to be crucial for our understanding of the construction of matter. Professor Steven Weinberg concludes the discussion of elementary particles by outlining the role that symmetry plays in our understanding of the laws of nature and elementary particles.

"An Astronomer's View of the Evolution of Man" by Sir Fred Hoyle provides a description of the origin of life and the limits of the universe's material resources. Sir Fred argues that an understanding of the nature of the universe's resources tells us a great deal about the long-term future of our species. He identifies some constraints that our chemical environment places on the use of natural resources, and he argues that nuclear energy, and not solar energy, is the most preferable alternative to resolving the current energy crisis.

Professor Stanley Jaki gives an historical account of what he considers to be the conceptual chaos threatening cosmology, and specifies the factors which he believes continue to plague even current scientific cosmological discourse. His argument in "The Chaos of Scientific Cosmology" has for its principal target a priori accounts of the cosmos with special reference to questions posed for cosmology by Gödel's incompleteness theorem and the concept of infinity.

In "The Place of Facts in a World of Values," Professor Hilary Putnam maintains that what is scientific is not coextensive with what is rational, and he suggests that there are many rational beliefs that cannot be tested scientifically. He argues that there are whole domains of real and objective fact that science tells us nothing about, not even that they exist. The three domains of fact considered by Professor Putnam are (1) objective values, (2) freedom, and (3) rationality itself.

DOUGLAS HUFF
OMER PREWETT

St. Peter, Minnesota
October 1978

Acknowledgments

The 1976 Nobel Conference was made possible by grants and personal gifts from the Charles K. Blandin Foundation, Blumberg Photo Sound Company, the Otto Bremer Foundation, Lloyd Engelsma, the General Mills Foundation, the Mankato Free Press, and Gustavus Adolphus College.

Appreciation also goes to the Nobel Foundation and to Consul General Per Olof Forshell, Royal Consulate General of Sweden, for his active support and participation in the conference.

As editors, we would like to express our appreciation to Beverly Lee for her remarkable transcription of the sessions and for her administrative assistance. We would also like to thank Elaine Brostrom for her excellent help in the final preparation of the manuscript.

D. H.
O. P.

Contents

The Nature of
the Physical Universe

What Is an Elementary Particle?

by

VICTOR F. WEISSKOPF

Institute Professor Emeritus
The Massachusetts Institute of Technology

Mr. President, ladies and gentlemen. I am greatly honored to be invited to address this conference. It shows me the tremendous interest in this part of the country for fundamental science.

I would like in this paper to talk about the concept of the elementary particle. The concept that matter consists of something fundamental, something elementary, is very deeply ingrained in our thinking. Obviously, we see an enormous variety of phenomena around us. However, within this variety we also observe recurrent properties. We observe, for example, the same materials in the same forms. We find constancy and regularity in the flow of events. People who have studied nature have concluded that this regularity is caused by something unchanging in that great flow of change around us. They have called that something an elementary particle.

This conception of an elementary particle has a very ancient history, going back at least to Democritus in ancient Greece. Typical of these conceptions is that of Newton. In his book *Opticks*, Newton wrote concerning elementary particles, the following:

> "All these things being considered, it seems probable to me that God in the Beginning formed Matter in solid, massy, hard, impenetrable, moveable Particles, of such Sizes and Figures, and with such other Properties, and in such Proportion to Space, as most conduced to the End for which he formed them; and that these primitive Particles being Solids, are incomparably harder than any porous Bodies compounded of them; even so very hard, as never to wear or to break into pieces; no ordinary Power being able to divide what God himself made one in the first Creation."[1]

And then he goes on and says,

> ". . . Changes of Corporeal Things are to be placed only in the various Separations and new Associations of Motions of these permanent Particles . . . ."[2]

These statements of Newton express clearly and distinctly the idea that matter is composed of elementary particles. This conception was corroborated by the chemists of the nineteenth century, who found that the tremendous variety of natural materials could be considered as being composed of only 92 different substances which we now call elements. These elements are composed of small particles called atoms. There are 92 different kinds of atoms, one for each element. During the nineteenth century each atom was thought to be indivisible, and this lack of divisibility was exactly what Newton had in mind when he said that primitive particles were "incomparably hard." In fact, the word "atom" itself comes from Greek, meaning nondivisible.

The fundamental idea of an incomparably hard elementary particle which could not be divided was not destined to last. In

1911, Rutherford made a decisive experiment that heralded a turn away from this fundamental idea. He showed that atoms do in fact contain smaller constituents. He literally took apart what according to Newton no ordinary power could divide. Furthermore, he showed us, what we now all know very well, that an atom is a planetary system with a nucleus in the middle, surrounded by scurrying electrons.

With Rutherford's discovery, physics and chemistry faced a crisis of extreme importance and deep significance. One aspect of this crisis was to understand how an atomic system that is a planetary system could be incomparably hard and unchanging. The stability of matter demands that atoms be hard and unchanging. Yet a planetary system is very easily changed if something disturbs it. It changes completely if one planetary system collides with another. For example, if our solar system were to collide with another, after the collision both systems would be quite different. Yet atomic systems do not behave this way. When two atomic systems collide, as happens very frequently in any gas, the atoms distort during the collision, but afterward they return to their original condition, with all the properties, such as shape and size, that they originally had. This process of stability through regeneration does not occur in planetary systems. Thus this aspect of the crisis was to understand how atoms could be planetary systems and yet at the same time be able to regenerate.

Another aspect of this crisis was the discovery that atomic systems had energy levels. That is, atoms can absorb and emit energy only in discrete amounts. For example, an atom in the lowest energy state can absorb a certain discrete amount of energy and become excited at its first level of excitation. It can give back this energy in the form of light and return to its original lower state. But it cannot accept only half as much energy and become somewhat less excited, or emit half as much energy and return only part way toward its original state. It is a case of all or nothing. This behavior contrasts strongly with that of a planetary system. A planetary system can absorb or emit energy in any amount it pleases. There is not a succession of energy levels characterizing a planetary system like there is in an atomic system.

A further difference involves the specific shape of atoms. Studies of the chemical bonding of elements suggest that the atoms of each element must have a definite specific shape, otherwise the manner in which different elements bond together and form molecules could not be understood. However, planetary systems do not have shapes that would account for the observed bonding.

It took physicists some two decades to resolve this crisis through the development of quantum mechanics. These decades were perhaps one of the most revolutionary periods in our quest for an understanding of nature. I cannot give you here an explanation of quantum mechanics. I would, however, like to point to the most important feature of quantum mechanics, the wave-particle duality. It is because of the wave-particle duality that we can reconcile the experiments of Rutherford, which demand a planetary model, with the stability and regeneration that occurs on the atomic level.

The wave-particle duality means that electrons exhibit wave-like properties under some experimental conditions, and particle-like properties under other conditions. The wave-like properties of an electron become particularly pronounced when the electron is confined. Figure 1 shows a symbolic presentation of the vibrational shapes that electron waves take when they are confined in a spherically symmetric region of space. In an atom, an electron is confined by the mere fact that the nucleus of the atom has a positive charge and the electron a negative charge. The coulomb attraction of these charges keeps the electron— whatever it is, wave or particle—within a finite distance. Because of this confinement, characteristic shapes, such as those shown in Figure 1, are produced. It is important to note that for any confined atomic system, the vibrations produce a unique

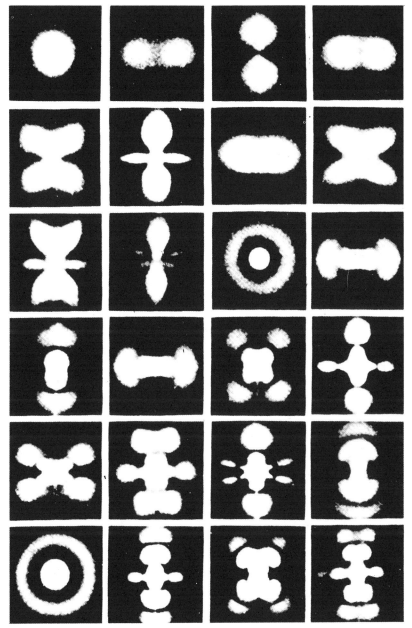

Figure 1 The vibrational shapes of electron waves confined to a spherical region of space.

5

series of characteristic shapes and absolutely no others. It is this unique series of atomic shapes that explains the atomic properties.

The regeneration of the atom arises because there is one and only one characteristic shape that has the lowest energy. It is this state that an atom normally occupies. Therefore, when an atom is perturbed because of a collision or for some other reason, when it loses its excess energy and returns to its lowest energy state, it must return eventually to the same characteristic shape that it started with. These results of quantum mechanics explain both regeneration, the uniqueness of the lowest wave pattern, and the existence of a lowest state.

It is important to note that there is a finite energy between the lowest state and the first excited state. This finite energy is of the order of several electron volts. This finite energy is very important because it is in terms of this energy that we understand the incomparable hardness of Newton. In Newtonian terms, this incomparable hardness means that an atom cannot be destroyed or undergo change. But this is exactly what is insured by this finite energy. As long as the surrounding conditions are not energetic enough, that is as long as they do not contain energies larger than this threshold energy, then the atom cannot but stay in the lowest state. Under these conditions the atom is stable, and has unchanging properties, namely the properties of the lowest state.

Our understanding of the incomparable hardness of atoms in terms of the finite energy difference between the lowest state and the first excited state, brings out an aspect of this hardness that was unknown to Newton. This aspect is what I like to call a conditional elementarity. Atoms act like elementary particles, that is they are incomparably hard, but only if they are in an environment that is not too energetic. When the environment contains energies much greater than the threshold energy between the ground state and the first excited state, then the atom has a large chance of becoming excited, and its properties change. That is, it no longer acts incomparably hard.

From the hardness of atomic particles, we turn now to their variety. There are some 92 different elements commonly found in nature. The various properties of these elements are accounted for in terms of variety of shapes that occur when an electron is confined. Look again at Figure 1, which illustrates some of the shapes that can arise when an electron is confined by a positive charge. In a many-electron atom, the electrons distribute themselves over shapes similar to these. In this distribution no two electrons can occur in a pattern with exactly the same shape. This restriction on the way that electrons can occur was found back in 1925 by Pauli and is known as the Pauli principle. This principle arises in a very general way in quantum mechanics. In an overly simple interpretation, it means that if one electron in a many-electron atom is in a shape corresponding to the lowest state in Figure 1, the next electron would have to be in a state corresponding roughly to the first excited state, the next electron in the next excited state, and so on. It is this principle that makes the variety of the elements we see in nature. If all of the electrons of a many-electron atom could congregate in a particular level, nature would be vastly different from what we observe.

Now let me repeat this in a different form, which shows essentially the same thing, namely that the wave patterns explain specific identity. If we have two atoms, say two hydrogen atoms, and the electrons form the same wave pattern in each, then the two atoms are identical. There is, so to speak, a qualitative difference in quantum mechanics between identical and nonidentical particles because of the discreteness of the states. All manifestations of a particular state are the same. Two different states are usually quite different. Since no quantum system can exist in a sort of in-between state, quantum systems must be either identical or in general quite different. In classical physics this is not the case. Two systems which are originally identical can be changed continuously and almost indistinguishably into different systems.

Now, I would like to discuss qualitatively how quantum

mechanics allows us to calculate the size and energies of an atom. In an atom there is a kind of equilibrium between two things: (1) the electric attraction of the positively charged nucleus and the negatively charged electrons, which keeps the electrons near the nucleus and (2) the expansion of the quantum mechanical wave. Quantum mechanical waves resist compression. The smaller the volume that the wave must occupy, the larger the energy necessary to maintain it in this volume. The wave patterns shown in Figure 1 are the result of a compromise, an equilibrium between these two effects. This equilibrium determines the size of the atom, and its energy levels. On the basis of these ideas, it can be shown that the radius of an atom is of the order of h^2/me^2, the Bohr radius, and the energy of the lowest atomic level is of the order of $-me^4/h^2$. In these formulas, m is the mass of the atom, e is the charge of the electron, and h is Planck's constant. The main point of all this is that quantum mechanics allows us to calculate the behavior of nature at the atomic level.

The dominant force in the structure of atoms is the electromagnetic force, that is, the force that charged particles exert on one another. Rutherford showed that the atom consists of a heavy, positively charged nucleus surrounded by electrons. The electrons are attracted to it, and quantum mechanics predicts what kinds of wave shapes will be produced. The heavy nucleus in the center plays the very small but decisive role of keeping the electrons around so that they are forced to take the interesting states shown in Figure 1. In some ways, quality is reduced to quantity, because if I tell you, "Here is an atom with fifteen electrons and a charge of plus 15 in its nucleus to keep these electrons around it," then all the atomic properties of this atom—its size, its color, the way it combines with other atoms, whether it is a solid or a gas—can be calculated.

I remember one of the most impressive experiences we had, my colleagues and I, during the dark days of World War II, when we were forced to work on the development of nuclear energy. We knew that an element which we called plutonium ought to exist. Nobody had ever seen it because it has a finite lifetime. All we knew was that the charge of its nucleus was 94 positive electron charges, and that it had 94 electrons around it. So we sat down and calculated its properties. The fact that this calculation could be done at all shows the tremendous power of quantum mechanics. We could predict that it must be a metal; that it must be brown colored; that it had a particular spectrum; that it had a certain hardness, and a certain elasticity. And when we had the first cubic inch of plutonium before us, it turned out that all of this was true.

The manner in which atoms bond chemically to form molecules can be explained, at least in principle, by the electromagnetic force together with quantum mechanics. I say in principle, because many of the processes are very complicated and we do not understand them in detail. But the principle of it is understood. We can picture the interaction leading to the chemical bond as the electric interaction of patterns such as those shown in Figure 1.

From our brief discussion of the chemical bond, we turn to molecular architecture. Every molecule has an internal structure. This structure is possible only because nuclei are heavy. They weigh several thousand times more than the electrons. It is a principle of quantum mechanics that the heavier a thing is the more localized it can be. Thus in a molecule the nuclei are localized because they are so heavy, while the much lighter electrons are not. These electrons are smeared out over the molecule and act as a sort of glue which keeps the molecule together. Thus in the case of water, for example, the nuclei of hydrogen and oxygen are localized very well, and determine the outlines of the molecule, while the electrons act as the glue to hold it together.

Now molecules are very much heavier than electrons, and it is natural to try to understand how this heaviness comes about. On the basis of Einstein's theory of relativity we know that mass is connected to energy through the celebrated formula $E = mc^2$. We can therefore understand the origin of the mass of a molecule if we understand what kind of energy causes it. We know that the electric interactions that give rise to the molecule do not have enough energy to explain the mass of that molecule. If, for example, I take a piece of rock and weigh it, and then calculate how much the electric energy in it weighs, it would turn out to be a ten-thousandth or less of the weight of the rock—practically nothing. Essentially all of the weight comes from energy associated with the nucleus, and not from the electrical interactions which bind the electrons to the nucleus, or the interactions that bind one atom to another in a molecule. So here is a big question mark, a question mark that points to the next story in the development of theoretical physics.

In order to understand the next step in the development of physics it is well to remember what I call the energy-size relation. It is a consequence of quantum mechanics that the smaller a system is, the higher the energy needed to get it out of its ground state, where it cannot change its properties, and into one of its excited states, where its properties are different. As I have already pointed out in connection with atomic systems, this excitation energy is the basis for conditional elementarity. If a particle is small then a large energy is necessary to excite it, and if that energy is not available the particle will act like an elementary particle. That is, it will act like a particle which Newton described as incomparably hard.

The next development in the search for the elementary particles of nature was the discovery of the neutron by Chadwick in the year 1932. He discovered that this was a particle contained in atomic nuclei. The importance of this discovery may be seen by looking at what the elementary particles of nature seemed to be before this discovery. Before this discovery, if I were asked to give the elementary particles of physics, I would have to give the unsatisfactory answer that they were electrons and nuclei. But the trouble is that all the 92 elements have different nuclei.

Each nuclei has a different charge on it. Hydrogen has one charge; oxygen, eight charges; uranium, 92 charges; and so on. In addition, we know that for a given charge there are so-called isotopes, so that in reality we have something like 250 different nuclei. Could they all be elementary particles? Of course people never really believed that they could. However, the nuclei are so very small, it takes high energy to excite them. So people believed that they had just not reached a high enough energy to find out what was going on inside. Thus when Chadwick in England discovered the neutron, it was clear that nuclei were not elementary particles, but that instead they were particles that are very hard to excite. It is now known that nuclei consist of neutrons, and another very similar kind of particle, protons. But these particles are so small, and the forces that keep them together so strong, that it takes about a million electron volts in order to get them excited. This is roughly 100,000 times more energy than is needed to excite atoms.

Now with the discovery of the constitution of nuclei a new world was opened, a new realm of phenomena. This world, which I call the nuclear world or the nuclear realm, deals with the phenomena that are connected with the internal structure of the nucleus. This internal structure is dormant in our environment, because the threshold for excitation is a million electron volts, and we just do not have that much energy available under ordinary conditions. An ordinary fire or even an explosion exchanges only a few electron volts between the atoms. Nuclear reactions require more than a million times as much energy. On earth energies of this kind can only be obtained by accelerators, by artificial radioactivity, or by cosmic rays.

So now we have new phenomena before us. What are the consequences? We can excite nuclei. This means that nuclei are really not elementary in the sense of being incomparably hard, unchangeable. Indeed, they are changeable, if only you have enough energy. A new kind of radiation emerges, which we interpret as radioactivity. The phenomena of fission and fusion occur. These were almost completely unknown before, although not quite completely, because there is natural radioactivity due to the fact that our earth was formed a few billion years ago under completely different conditions than are present now. In fact, probably several billion years ago there was a star explosion. At that time energy exchanges of a million electron volts were easy to be had. The radioactivity observed today is just the last embers of this cosmic fire in which the material of the earth was created. Clearly we are entering a new realm of nature, the exploration of which is one of the most significant developments of the physics of this century.

One phenomenon which must be the subject of this exploration is the nuclear force. Clearly if the nucleus is composed of protons and neutrons, and the neutrons are not charged, the electromagnetic force between charged particles cannot hold them together. There must, therefore, be a new force, a very strong force, to bind these particles together. That force is called the nuclear force. Figure 2 shows the potential curve for the nuclear force. This force is attractive when the curve goes down and it is repulsive when the curve goes up. As can be seen from the curve, this force is attractive at a distance of two or three units. This is a very small distance since the units are 10^{-13} cm, that is 1/100,000 of an atomic size. The energy scale is in units of 10 million electron volts.

Now when we study the structure of the nuclear world, we find that it is possible for neutrons in the nucleus to turn into protons. This means that the neutrons and protons are really just two forms of the existence of a single particle called a nucleon. The neutron state of this particle can change into the proton state and vice versa with the emission of a radiation which we call lepton pairs. The word "lepton" is a common name for electrons and neutrinos. The discovery that the neutron and proton are two states of the same particle introduced a new symmetry into physics. This symmetry is the replacement of neutrons by protons and vice versa. If neutrons and protons are

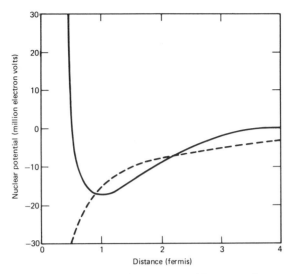

Figure 2   A sketch of the potential of the nuclear force as a function of the distance (not quantitatively correct, as it depends somewhat on spin and symmetry). For comparison, the dotted curve represents the attraction of two opposite charges.

really just names for two different charge states of the same particle, the nuclear forces should be unchanged when they are interchanged, since these forces are not affected by charge.

Figure 3 gives a nuclear energy spectrum of two nuclei which differ by having the protons exchanged for neutrons and vice versa. They are called mirror nuclei. Notice that they are not very different. This shows that the nuclear force does not depend on the charge.

Figure 4 gives another nuclear spectrum. The energy scale is a million electron volts. Here we see illustrated a new phenomenon. In the atomic realm when an atom changes states, light quanta are either absorbed or emitted. Here the same basic thing happens except when the nucleus changes states. Not

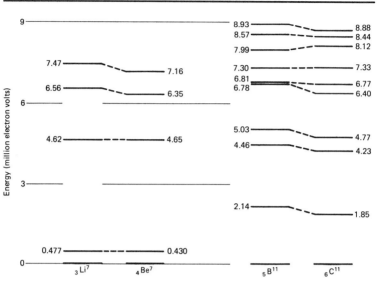

Figure 3   Nuclear spectra for "mirror" nuclei. Spectra of mirror nuclei are very similar because the nuclear force does not distinguish neutrons from protons. Two nuclei form a mirror pair if they contain the same total number of protons and neutrons and if the number of protons in one nucleus equals the number of neutrons in the other. Thus $_3Li^7$ (lithium) and $_4Be^7$ (beryllium) form one mirror pair and $_5B^{11}$ (boron) and $_6C^{11}$ (carbon) form another. The subscript indicates the number of protons in the nucleus, and the superscript indicates the total number of protons and neutrons. (From *The Three Spectroscopies* by Victor F. Weisskopf. Copyright © 1968 by *Scientific American, Inc.* All rights reserved.)

only are light quanta, called gamma rays, emitted or absorbed, but also lepton pairs. These lepton pairs cause the charge of the nucleus to change.

At this point it is interesting to note a difference between the nuclear and the atomic realm. In the atomic realm there is a heavy nucleus and light electrons. This situation is almost characterizable as an authoritarian regime. The heavy, highly charged nucleus decides what happens. By the same analogy, the nucleus can be characterized as a democratic regime. The protons and the neutrons are practically the same weight, and exert the same nuclear force.

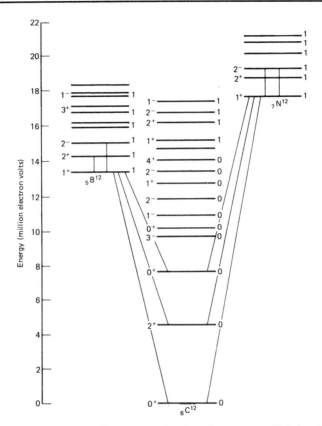

Figure 4 The energy level spectrum of a 12 nucleon system. This is a typical nuclear spectrum. The numbers on the left of each level give the angular momentum *J* and parity, whereas the numbers on the right give the isotopic spin. The connecting lines are the most important transitions; the vertical ones are electromagnetic, while the skew ones involve lepton pair emissions.

We have now traced the development of the idea of an elementary particle up roughly to the 1930s. At this time physicists lived in what could be described as a fool's paradise. Physicists thought they had found all the elementary particles. These were thought to be the electrons, protons, neutrons, neutrinos, and light quanta. Why was it a fool's paradise? Well, it was a fool's paradise because it turned out it was not so.

Well, how did it all turn out? To explain this, I will start by talking about what I like to call the quantum ladder. Molecules consist of atoms. Atoms consist of nuclei and electrons. Electrons do not yet seem to be composite, but nuclei consist of nucleons, that is protons and neutrons. Now what about the nucleon? What does it consist of?

In answering this question we already have a model which may act as our guide in atomic systems. Atoms attract each other to form molecules, just as nucleons attract each other to form nuclei. So by comparing the force between atoms with that between nucleons we may get a clue which would help us understand this latter force. Now when we look at the force of chemical bonding, illustrated by means of the potential function, we find that it has a number of similarities to the nuclear potential shown in Figure 2. Atoms as well as nucleons are attracted at a certain distance and then repelled at closer distances. There are, however, certain differences. The chemical force is much weaker. The energy units for the chemical force are only electron volts and not millions of electron volts as in the case of the nuclear force. The distance units are much larger. They are angstroms and not fermis; that means 10^{-8} cm instead of 10^{-13} cm. Nevertheless, the potential functions of the two forces are strikingly similar.

Since the forces seem to be similar, it may be the case that we can understand the nuclear force in the same way that we understand the chemical force. I have mentioned earlier that the chemical force is well understood. The chemical force is nothing else than the electrical force plus quantum mechanics. Now since the nucleon force has the same basic shape as the chemical bonding force, there is hope that we can understand it in a similar fashion, that is, in terms of a nucleon internal structure analogous to the internal structure of the atom. Of course the analogy is not exact. The nucleon force is not identical to the chemical force in all respects; only in general outline. The nucleon force is a complicated force depending in detail on such things as isotopic spin and other things I have not mentioned. But the analogy is good enough to give us hope that a search for the internal structure of a nucleon would be profitable.

The first experimental evidence of the internal structure of the nucleon came when in 1952 Fermi and his collaborators in Chicago were able to excite a proton to its first excited state. This result led towards a new world, a third realm of phenomena which I like to call the subnuclear world.

I would like to give here a survey of the strange unexpected phenomena, dormant under normal conditions, which we find here. Because protons are smaller than nuclei, the energy size relation of quantum mechanics which we have discussed before dictates that the energy we need to excite these particles will be larger than that needed to excite the nucleus. Mr. Fermi and his colleagues needed a cyclotron of 400 million electron volts, 400 times more than the usual threshold in nuclei, in order to obtain the first excited state of the proton. Once enough energy is available to excite these nucleons what does one find? Well, one finds excited protons and neutrons, of course, and then something new: mesons of all kinds, heavy electrons, a new neutrino, antimatter, and the production of other particles in lavish numbers.

The next four figures illustrate some of these discoveries. Figure 5 gives the nucleon spectrum as it was known about ten years after Fermi's first discovery. Fermi discovered the first excited state, which is the lowest one in the second column. All the other states have been discovered since then. In fact these data are about 10 years old; there are many, many more levels now known than are shown here. Notice that the energy scale is in gigaelectron volts, which means a billion electron volts. This is a much higher energy than we have needed at the atomic, or nuclear levels. A lot of transitions are possible here.

Figure 6 illustrates some of these subnuclear transitions in

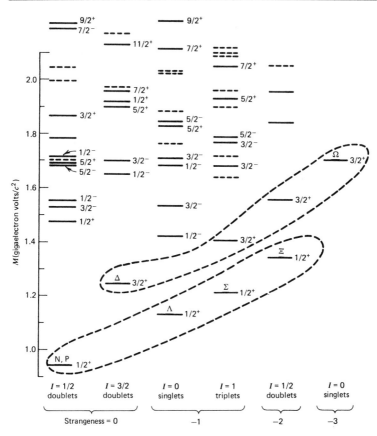

Figure 5   The energy level spectrum of the nucleon. The numbers on the right of the energy levels indicate the angular momentum quantum number and the parity. The broken lines encircle the SU₃ octet (lower one) and the SU₃ decuplet (upper one).

comparison with the transitions seen at the atomic and nuclear levels. We have gamma rays (light quanta) and electron pairs as in the nuclei, and new things, π and K mesons.

It soon was found that the π and K mesons were not the only mesons that existed. There are many more. Figure 7 is a plot of the whole meson spectrum, of which the π and K mesons are only a sample.

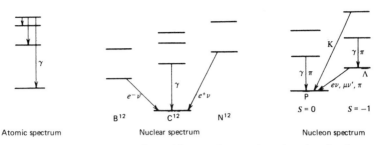

Figure 6   A comparison of transitions at the atomic and nuclear levels.

The mesons are rather ephemeral entities. They do not exist very long. In fact, they decay in about a billionth of a second or less into less exotic particles.

Figure 8 shows how some of the different mesons decay. Notice that π and K mesons decay into gamma rays (that is, light), and into lepton pairs. It turns out that the electron appears in two forms, a normal electron and a heavy electron. The heavy electron has historically been known as a muon. It is unstable, ultimately decaying as shown in Figure 8.

The presence of a muon leads us into a completely new world again. But before I say more about it, I must say something about antimatter.

In the 1920s, in fact at the time I was a student, physicists were surprised that atoms had positively charged nuclei and negatively charged electrons. They wondered why there were not atoms in which the nucleus was negatively charged and surrounded by positively charged electrons. "Why is nature so asymmetrical?," they asked. Some physicists thought that nature was just that way—unsymmetrical. Then came a development which I believe is one of the most dramatic developments of the intellectual penetration of nature by man. Paul Dirac, in England, was trying to adjust quantum mechanics to special relativity. This was not easy. In the course of doing so, he came to an equation that, and now I am almost quoting him from a talk he gave two years ago, scared him because it was so infi-

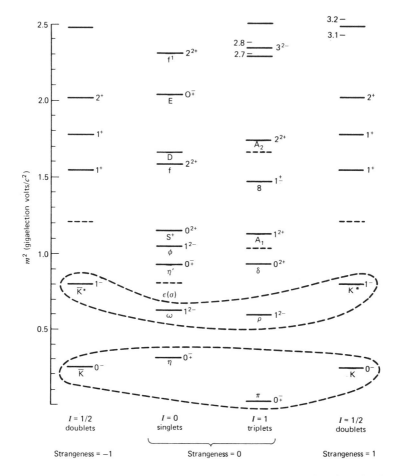

Figure 7 The spectrum of mesons. The numbers on the right of the levels indicate the angular momentum quantum number, the parity sign, and, in some cases, the charge-conjugation sign. The broken lines encircle two SU₃ octets, the lower one with $J = 0$, and the upper one with $J = 1$.

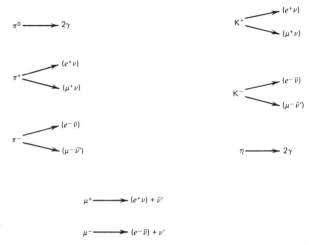

Figure 8   Decay schemes of selected mesons. In this figure γ means a photon, $e^\pm$ means a positive or negative electron; ν and ν' are neutrinos, $\bar{\nu}$ and $\bar{\nu}'$ are anti-neutrinos, and μ is a heavy electron (muon).

nitely rich in information. This equation showed why the electron has a spin, why it has a magnetic moment. But it also contained the statement, if you read the equation right, that if a negative electron exists, then a positive electron must also exist; if a positive proton exists, then a negative proton must also exist. In other words this equation has the symmetry built into it, that if a certain particle exists, so must another particle, its so-called antiparticle. Dirac, according to his own words, was so scared by this that he just did not believe it. He thought that there was something wrong, that he had interpreted something wrongly, because nobody had, at that time, seen a positive electron or a negative proton. Indeed, he first interpreted the equation wrongly just because he could not believe that the correct interpretation could be true. However, only a few years later, in 1932, Anderson and Neddermeyer discovered the positive electron. This experiment and many subsequent ones have confirmed the existence of the particle antiparticle symmetry described by the Dirac equation.

Now the particle antiparticle symmetry is connected with a strange phenomenon: the production and annihilation of subatomic particles. One aspect of this phenomenon is the annihilation of particle antiparticle pairs to other forms of energy, that is, to light, or to mesons. For example, a proton and an antiproton together can annihilate producing light or mesons. Another aspect of this phenomenon is the creation of particle antiparticle pairs from pure energy, that is, light. For example, light can be transformed into electron-positron pairs, or proton-antiproton pairs. These interactions explain why we do not see antimatter under ordinary conditions. If antimatter were to come into our world, it would annihilate with ordinary matter to produce light or mesons. So under ordinary circumstances we see only the ordinary matter which is left after all the antimatter has been annihilated. Either of the two kinds of matter can be stable in any one region of space as long as none of its antimatter is present at the same time. If both kinds are present, they will interact and annihilate each other.

Now the production of antimatter poses a great difficulty for the physicist in the subnuclear realm. This difficulty arises because we are dealing with extremely high energy. At these energies pairs of particles and antiparticles can be produced. A lot of energy is needed for particle-antiparticle production because enough energy must be available to produce the rest mass of each particle. Thus to produce an electron-positron pair, twice the rest mass energy of the electron is needed. This is a large amount. Under ordinary conditions in the atomic world we just do not have the energies available to produce these particles. That is why antimatter is produced only under exceptional conditions in the atomic or the nuclear world. But when we come to the subnuclear world then the energy is indeed great enough. That places us in a very strange situation. I shall describe the difficulty with an analogy. Take a couple of good

Swiss watches. If you want to know what is inside them, one method of finding out is to hurl them together so strongly that they break open, revealing a lot of wheels, springs, and so on. This is basically the method physicists have used to find out what is inside of atoms and nuclei. But this method does not work with nucleons. If two nucleons are hurled together strongly enough to see what is inside of them, not only is the internal structure of the nucleons exposed, but also a lot of particles and antiparticles are created. The difficulty comes in deciding what particles come from the internal structure of the nucleon and what are merely being created because of the high energies which are available. (Now indeed this difficulty is true even for the Swiss watches, because when you hurl them together strongly enough to break them, sparks fly from them. No one thinks of the light emitted as being in the original watches!)

The data from our experiments with subnuclear particles are given in Figures 5 and 7. Figure 5 gives some of the many ways the proton and the neutron can exist; Figure 7 shows the mesons. The question that these data pose is, of course, how to understand the underlying structures from which this plethora of particles arises. It reminds me of the situation when we had 92 nuclei, a plethora of nuclei. Then we found out that there really are only two kinds of nuclear particles, protons and neutrons, and all of the 92 nuclei are just combinations of these. And now the development discussed by Dr. Murray Gell-Mann is that all of these mesons and baryon states are nothing else, hopefully, than a combination of three types of quarks. Figure 9 identifies various types of quarks. In recent years more states have been discovered which require the existence of more than three quarks. I will not discuss them here because Dr. Gell-Mann will discuss them thoroughly.

I would like now to summarize the developments about which we have been talking. If we look at a piece of matter, say a piece of copper, we notice it is formed of regularly spaced atoms. If we then look at a single atom, we see a nucleus and an

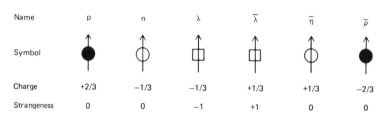

Name	p	n	λ	$\overline{\lambda}$	$\overline{\eta}$	$\overline{p}$
Symbol						
Charge	+2/3	−1/3	−1/3	+1/3	+1/3	−2/3
Strangeness	0	0	−1	+1	0	0

Figure 9   Three quarks, and their antiparticles (right), have been proposed as the hypothetical building blocks of baryons and mesons. Unlike all particles discovered so far, quarks would carry less than a whole unit of the charge of the electron. The lambda (λ) quark is provided with a negative unit of strangeness. Each quark and antiquark carries half a unit of spin, the direction of which is shown by an arrow. As in the world of known particles the hypothetical antiquark mirrors the properties of the quark. (From *The Three Spectroscopies* by Victor F. Weisskopf. Copyright © 1968 by *Scientific American, Inc.* All rights reserved.)

electron. Then if we look at one nucleon, we suspect that it is composed of quarks. (For myself I am not quite sure of this latter point. Maybe Murray will convince me of it.) Just as the electrons are bound to the atoms by the electric force, we expect that there is a strong force which keeps the quarks together. This force may explain the nuclear force in the same way as the electric force explains the chemical force.

Now I want to answer the title question, what is an elementary particle? The answer depends on what you mean by the question. It involves what I have called conditional elementarity. If you mean by that question that you are dealing with energies of less than about a tenth of an electron volt, then molecules such as we find in the air around us are elementary particles. At these energies, molecules are unchanging and appear infinitely hard, and collisions will not change them. If you mean higher energies, then atoms and photons are the elementary particles. If still higher energies are meant, then nuclei, electrons, and photons are elementary particles. If you go into energies in the millions of electron volts range, then the elementary particles are protons, neutrons, electrons, neutrinos, and photons. Finally, if you go into the subnuclear realm you have quarks, electrons, neutrons and protons, heavy electrons, neutrinos, light quanta, and with a lot of question marks, maybe some other quanta such as gluons.

I would like to end with a quotation from Newton. In the same book *Opticks* which I quoted before, Newton wrote the following farsighted statement:

> "Now the smallest Particles of Matter may cohere by the strongest Attractions, [perhaps quarks?] and compose bigger Particles of weaker Virtue [nucleons?]; and many of these may cohere and compose bigger Particles [atoms?] whose Virtue is still weaker and so on for the diverse Successions until the Progression end in the biggest Particles on which the Operations in Chymistry, and the colours of natural Bodies depend, and which by cohering compose Bodies of sensible Magnitude."[3]

And then later he says,

> "There are therefore Agents in Nature able to make the Particles of Bodies stick together by very strong Attractions. And it is the business of experimental Philosophy to find them out."[4]

NOTES AND REFERENCES

1. Isaac Newton, *Opticks*, 4th ed., 1730 (reprinted by Dover, New York, 1952), p. 400.
2. Ibid., p. 400.
3. Ibid., p. 394.
4. Ibid., p. 394.

What Are the Building Blocks of Matter?

by

MURRAY GELL-MANN
Robert A. Millikan Professor of Physics
California Institute of Technology

It is a pleasure to be here today, and to be honored by such nice introductions. I have had two of them so far. The one at the first convocation mentioned that I received my doctor's degree in physics at the Massachusetts Institute of Technology a little more than 25 years ago. It did not mention that my thesis supervisor and teacher at MIT was none other than Victor Weisskopf, whom you heard an hour ago talking about physics. I found working with him a magnificent experience and we have been friends ever since. A few years ago, living near each other in the French Jura, we even arranged meetings in the woods, at which we would sit on the opposite ends of a log and talk about physics while our dogs frolicked under the spruce trees. Because of this old and very friendly connection with Professor Weisskopf I will forgive him for (1) giving half of my talk, and (2) telling you not to believe the rest. I hope you will in turn forgive me if there is some overlap between the first few remarks and the content of his speech.

I must apologize for speaking somewhat technically today, but I think it is difficult to talk about particle physics without being a little bit technical. I must also apologize for emphasizing theory in my presentation. I am of course a theorist. I work with pencil, paper, and wastebasket, just like Professor Weisskopf. Nevertheless I do not want to give the impression that the field progresses by theoretical work alone. Experiment and theory are equal partners in the struggle to find out more and more about the basic laws of physics. There is not enough time today to discuss the experimental evidence for the theoretical work that I shall present. But in principle one should certainly present them together.

Particle physics investigates the building blocks of all matter everywhere in the universe. As far as one can tell from signals arriving from the most distant regions, matter has exactly the same properties everywhere. These properties, including the forces (or interactions) among the particles, constitute the fundamental laws of physics on the microscopic scale. To these we must adjoin the boundary conditions of the whole universe, including its existence some 15 billion years ago as a tiny expanding ball of dense matter. Taken together these microscopic and macroscopic principles then constitute the fundamentals of physics that underlie all of astronomy, chemistry, geology, and so on, as well as the various branches of physics itself. We are all familiar with one of the building blocks, the elementary particle called the electron. Electrons, as you know, form the outer portions of atoms and play a leading role in chemistry. As for forces, we are all familiar with at least two of them, electromagnetism and gravitation. Both forces vary in the same way with distance, at least approximately. But between two particles, two electrons for instance, the gravitational force is some 10^{42} times weaker than the electrical force. So particle physicists have tended not to pay too much attention in practice to gravitation. Nowadays that is changing as we learn more about how to integrate gravitational theory with the theory of the other forces, but I shall not mention it very prominently in this paper.

As Professor Weisskopf pointed out, all of modern physics is governed by that magnificent and confusing discipline called quantum mechanics. The philosophical interpretation of quantum mechanics is still probably not complete, but operationally quantum mechanics is in perfect shape. (The fact that an adequate philosophical presentation has been so long delayed is no doubt caused by the fact that Niels Bohr brainwashed a whole generation of theorists into thinking that the job was done 50 years ago.) Anyway, quantum mechanics itself has survived all tests and there is every reason to believe it to be exactly correct. But it is not in itself a detailed theory of forces or interactions. It is a framework into which any correct theory must fit. In our work in elementary particle theory we assume it as a basic principle. We assume also in our work two other basic principles: relativity, developed by Einstein in the early years of this century, and causality, a very simple principle to which we all subscribe, and which states that a cause precedes its effect, or equivalently, signals travel forward in time. Any theory that obeys all these principles is called a local quantum field theory and all our serious attempts today to describe fundamental microscopic reality in physics are in the domain of local quantum field theory.

The first local quantum field theory started in 1929, the year I was born. The theory formulated in that year has worked ever since, as more and more and more accurate experiments have tested it to more and more decimal places. It is called quantum electrodynamics, abbreviated as QED, and it is the quantum mechanical version of electromagnetic theory, that is, the theory of electrons and photons. Electromagnetism, like all

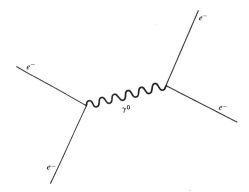

Figure 1   Feynman diagram of two electrons (labeled e^-) exchanging a photon (labeled γ^0).

forces in quantum field theory, is carried by a quantum. It is called the photon in the case of electromagnetism, from the Greek word for light, since light waves are a kind of electromagnetic wave.

Now in QED you can draw the little picture shown in Figure 1. These pictures were invented by my colleague, Dick Feynman, and they help us to describe how a force occurs in a quantum field theory. Two electrons are shown; they exchange, in the sense of quantum mechanics, a photon. A photon is electrically neutral and it is written as gamma with a little zero above it to indicate this fact. The electrons have one unit of negative electric charge. The exchange of a "virtual" photon between them gives rise to the electromagnetic force, which causes them to repel each other, the charges having the same sign.

But QED, no matter how beautifully it has worked so far, cannot be a perfect theory forever, because we know it is not complete. It does not include all sorts of other things that exist in the world besides electrons and positrons and photons. So we must generalize this theory to include all the other phenomena. From Dr. Weisskopf's paper, we can surmise what some of the other objects are that we have to mention in the course of our work. In the atom there are not only the outer layers that do consist of electrons, but also the particles in the nucleus. The nucleus seems to be made of electrically neutral neutrons and electrically positive protons. The chemistry is of course determined by the number of electrons, which in turn equals the number of protons in the nucleus, so that the atom is balanced in charge. (The "isotope" is determined by how many neutrons there are in the nucleus in addition to the protons.)

Now it is already clear from the previous paper that the nonsense in textbooks about neutrons and protons being elementary should not be believed. Textbooks as usual are out of date. You heard a lot of evidence that the neutron and proton are in fact composite. Just as molecules are made up out of atoms, atoms are made up out of nuclei and electrons, and nuclei are made up out of neutrons and protons, so the neutron and proton are made up out of quarks. All of this was indicated in Professor Weisskopf's paper. He said not to believe in the quarks, but he confided to me that actually he does believe in them fully, but just does not want to admit it in public.

Now all of these things—the molecules, the atoms, the nuclei, and then the neutron and proton—exhibit structure, as you know from the previous paper. They have excited energy levels. Furthermore, and this is very important, not only do they have excited energy levels but many of these excited energy levels show that the constituents revolve around one another. That is, the whole system exhibits angular momentum, which we can excite at will to higher and higher values. Angular momentum of this kind is a characteristic of composite systems. All of the systems we mentioned exhibit this behavior, and they are clearly composite.

Now the idea that neutrons and protons are made of quarks is quite old, more than 13 years old, and it has been confirmed by a great number of experiments and a large amount of theoretical work. According to our prescription, the neutral neutron and positively charged proton are each, roughly speaking, made of three quarks bound together. The neutron is made of a u quark and two d quarks. The u quark has an electrical charge of $+\frac{2}{3}$ in units such that the proton has charge $+1$ and the electron has charge -1. The d quark has charge $-\frac{1}{3}$. The neutral neutron is made up of one u and two d's, and the charges add up to $\frac{2}{3} - \frac{1}{3} - \frac{1}{3} = 0$. The proton with its $+1$ charge is made up of two u's and one d, since $\frac{2}{3} + \frac{2}{3} - \frac{1}{3} = 1$. Therefore, we represent the neutron n^0 by $u^{+2/3} d^{-1/3} d^{-1/3}$ and the proton p^+ by $u^{+2/3} u^{+2/3} d^{-1/3}$ to show their composition in terms of quarks. The u and d are called flavors of the quarks and as we shall see there are other flavors known as well. This nomenclature is not connected with actual flavor is in sweet and sour, and so on, but it is a nice analogy. We amuse ourselves during the long hours of doing integrals by using these names.

Now the excited states of neutrons and protons include not only neutral states made of $u^{+2/3} d^{-1/3} d^{-1/3}$ and positive states made of $u^{+2/3} u^{+2/3} d^{-1/3}$ but also, as you might expect, negatively charged excited states made up of $d^{-1/3} d^{-1/3} d^{-1/3}$ ($-\frac{1}{3}, -\frac{1}{3}, -\frac{1}{3}$ adding up to -1), and also states of charge $+2$ like $u^{+2/3} u^{+2/3} u^{+2/3}$ ($+\frac{2}{3}, +\frac{2}{3}, +\frac{2}{3}$ giving $+2$).

The quarks have not only flavor but also another property, which we playfully call color. It has nothing to do with real color, of course, but we call it that. I cannot blame other people for all of these pet names, because I made up most of them myself, although not "flavor," which was invented by Yōichirō Nambu. Anyway, the names of the colors are red, green, and blue. (The names refer to a somewhat oversimplified theory of human color vision in terms of three primary colors; note white is then a kind of average of all three. You could also use red, white, and blue, if you feel patriotic, in the United States or France or Britain or Holland or Norway. But you cannot use the colors of the flag everywhere. In Austria, for example, the flag is degenerate: red, white, and red.) Anyway, the prescription is the following: a neutron or proton is made of one quark of each color in a particular mathematical pattern such that color averages out. Remember there are three quarks and three colors and you put them together in a specific way with one quark of each color. You can think of the neutron and proton as being white, if you like, with the three primary colors averaged out; all the excited states of neutron and proton likewise. In fact, all of the strongly interacting particles that anyone has seen are white in this sense. They all are made up of quarks in such patterns that the color of the quark is made to average out completely.

Now we have a very definite mathematical theory of the interaction of these quarks with one another. This theory is rather similar to QED. The quarks are bound together by the exchange of quanta called gluons. The force between the quarks, in other words, comes from exchanging these gluons just the way the force between two electrons in electromagnetism comes from exchanging photons. Figure 2 shows a Feynman diagram of a couple of *u* quarks exchanging a gluon.

The gluons do not pay any attention to flavor; they are coupled equally to the different quark flavors. There is not just one type of gluon; there are different gluons for different color situations. We illustrate these different gluons by drawing

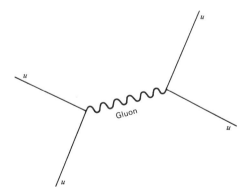

Figure 2   Feynman diagram of two *u* quarks exchanging a gluon.

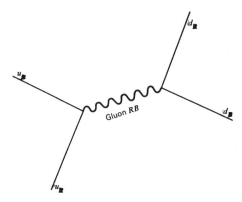

Figure 3   The interaction of a red u quark and a blue d quark through the exchange of a red-blue gluon. This interaction changes the red u quark into a blue u quark and the blue d quark into a red d quark.

Feynman diagrams showing color. Suppose we have a force between a u quark and a d quark, illustrated in Figure 3. In the course of an interaction, the u quark can turn from a red one into a blue one, and the d quark can turn from a blue one into a red one, as a result of exchanging what we might call a red-blue gluon. The flavor matters not at all here. I can replace the u quark by a d quark and the diagram comes out exactly the same. As long as the color pattern is the same you always exchange a red-blue gluon. In Figure 4 we have a blue-green gluon; here I happened to attach the label d to each of the quarks. A blue quark is turning into a green quark on one side, while a green quark is turning into a blue quark on the other side.

The definite quantum field theory that we propose for quarks and gluons, analogous to the QED that has been so successful, is called quantum chromodynamics (QCD), or quantum color dynamics, if you want to translate chromo into English. In QED we have electrons, which are electrically charged, and we have photons, which are neutral, and the electrons exchange photons in order to give the electromagnetic force. In QCD, we have

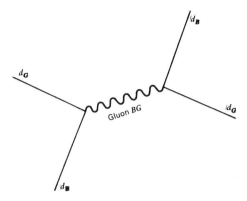

Figure 4   The interaction of a blue *d* quark and a green *d* quark through the exchange of a blue-green gluon. The blue *d* quark changes into a green *d* quark and the green *d* quark changes into a blue one.

various flavors of quarks each coming in three colors, red, green, and blue, and the force comes from exchanging these colored gluons. Like photons the gluons are electrically neutral. But they are not color neutral. The gluons have color in the color theory, whereas in the electrical theory the photons do not have any electric charge, and that makes a huge difference between the two theories. It means the equations, which otherwise look rather similar, have totally different properties. In the case of QED we can solve the equations at large distances but not at very small distances, whereas for QCD we can solve the equations easily at small distances but we have so far not solved them convincingly at very large distances.

Many of us think we can guess what the solutions will look like at large distances. (I suspect that within a few years people will know.) We think that the color force does not decrease at large distances.

We know that at very small distances the effective color charge increases with increasing distance between quarks. What we conjecture is that at very large distances (greater than 10^{-13} centimeters) the effective color charge grows so rapidly that the actual force between quarks does not decrease, an unheard-of situation in physics. The colored quark and the colorful gluon are then confined by the attractive force between colors, which refuses to decrease at large distances. Colored objects are permanently trapped inside observable objects and cannot get out. "White" nuclear objects like the neutron and proton, which are made up of quarks and the gluons that bind them together, can never eject these colored particles because of the way the color force behaves. We can never get these colored particles out singly to examine them; they can never be directly detected.

This is what most of us believe to be the case, although there are a few theorists who doubt it. We must investigate further what the equations say and also what the experimental evidence supports. People have searched very hard for isolated, directly observable quarks and have never seen any. This search will continue, but most of us believe that it is completely hopeless, that they will never be detected because the forces are such as to trap them permanently along with their gluons.

The idea that quarks are not directly observable does not mean that the quark and gluon theory is an idle speculation that can never be checked by experiment, like the question of how many angels can dance on a head of a pin. We can do high energy collision experiments on neutrons and protons and all their friends and we can see if the neutrons and protons act as if they were made up of colored quarks held together by colorful gluons. Everything we have seen so far is in agreement with that hypothesis. However, more calculation and more experiments will be helpful in making sure. Then, as Professor Weisskopf suggested, we can explain the nuclear force that binds the neutrons and protons together in the atomic nucleus as a secondary effect of the basic quark interaction that comes through the exchange of gluons. In the same way, the binding of atoms in a molecule is known to be a secondary effect of the electromagnetic force acting among the electrons and nuclei.

Now the electron does not feel the nuclear force. In fact in very heavy atoms the innermost electrons spend a great deal of their time inside the nucleus. They never feel the nuclear force that the neutrons and protons feel so strongly. They feel only the electrical force of the electrically charged protons. Since the electron does not have any nuclear force, that means it does not have any color. Lacking color, it is not coupled to the gluon, that is, it is not coupled to the color force and that is why it is not coupled to the strong nuclear interaction in the way the neutrons and protons are. However, the other property, flavor, which distinguishes neutrons from protons and which distinguishes u quarks from d quarks, does have an analog in the case of the electron. Another flavor of electron, we can say, is the electron neutrino. It is indicated by the symbol ν_e^0, which is the Greek letter ν with the subscript e because it is a friend of the electron, and with the superscript 0 meaning that it is electrically neutral. This particle has very little interaction with anything because it does not have a nuclear interaction nor does it have an electromagnetic one. It can pass right through huge amounts of matter with only the slightest probability of interaction. If we succeed someday in detecting the neutrinos from the sun (which for some puzzling reason we have not so far succeeded in doing), we could detect them in the daytime coming from above and at night coming from below because they do not pay much attention to the earth.

Now the electron neutrino and its friend the electron can be converted into each other by another force called the weak force. For instance Figure 5 shows a reaction which takes place through the weak force. The electron neutrino strikes a neutron, both of them being neutral, and the neutrino turns into an electron while the neutron turns into a proton. It has been known for a good many years that this can happen. More basically, we can describe this interaction in terms of quarks. Here the electron neutrino strikes a d quark with charge $-\frac{1}{3}$ resulting in the neutrino turning into its friend the electron while the

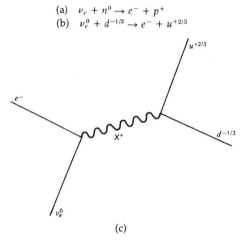

(a) $\nu_e + n^0 \rightarrow e^- + p^+$
(b) $\nu_e^0 + d^{-1/3} \rightarrow e^- + u^{+2/3}$

(c)

Figure 5 The interaction of an electron neutron (ν_e^0) and a neutron (n^0) to produce an electron (e^-) and a proton (p^+): (a) phenomenological statement of the interaction; (b) the same interaction interpreted as the neutrino acting on the $d^{-1/3}$ quark in a neutron and converting it into a $u^{+2/3}$ quark of the proton; (c) Feynman diagram of the interaction given in (b); note that the reaction proceeds by the interchange of a hypothetical x^{+1} particle.

d quark turns into its friend the u quark. The charge of the d increases by +1, while the charge of the neutrino decreases by 1, so the total charge is the same on both sides. Figure 5 shows a Feynman diagram of this process, which takes place through the exchange of a new kind of quantum which, when I invented it, I called X, but which some people seem to call W instead. (I am still pumping for X, the correct name.) The neutrino emits an X^+ particle, and turns into an electron, while the d quark absorbs the X^+ quantum, and turns into a u. "Absorbs" and "emits" are used as usual in the Pickwickian sense of quantum mechanics.

In addition to the quantum X^+, there also is a similar quantum X^-. These electrically charged particles are heavy, and we predict that they will be seen in very high energy experiments within a few years. We do not even need fantastically expensive new equipment in order to see them. In fact it is just barely conceivable that they can be detected with present equipment. More likely we will need a storage ring to go with the existing very high energy accelerator at Fermilab near Chicago or the one at CERN near Geneva. Such a storage ring does not increase the cost of the whole installation by a gigantic factor, the way many previous new developments have, and it should make possible the detection of these so-called intermediate bosons for the weak interaction. It is very important to find them, so that we can be sure that we are still on the right track in talking this way about quanta always being associated with forces.

Now the electromagnetic and weak forces are flavor forces. They operate on the flavor variables. The electric charge varies with flavor. The weak force actually flips one flavor to another one. Right now a unified quantum flavor dynamics is emerging, to describe in a unified manner all of these flavor forces, including electromagnetism. Unified flavor dynamics is a generalization of QED that includes it, but also includes that other flavor force, the weak force. Steven Weinberg, who presents the next paper, has played an important role in helping to formulate this quantum flavor dynamics, which is not yet finally written down, unlike QED. We have not fixed on a definite candidate for this theory, but we are looking for one very hard.

So far the effort to find a unified quantum flavor dynamics has had several nice consequences; one is that it has successfully predicted a new flavor force that is essential in order to make a unified description of the flavor forces. You can call it a "neutral weak interaction" if you want, or a "noncharge-exchange weak interaction" or even a "neutral-current weak interaction." By means of this force a neutrino can simply scatter from a proton, as shown in Figure 6. A neutrino plus a proton gives a neutrino plus a proton; and similarly for neutrons. More fundamentally a neutrino scatters from a quark as is

(a) $\nu_e^0 + p^+ \rightarrow \nu_e^0 + p^+$
(b) $\nu_e^0 + n^0 \rightarrow \nu_e^0 + n^0$
(c) $\nu_e^0 + u^{+2/3} \rightarrow \nu_e^0 + u^{+2/3}$
(d) $\nu_e^0 + d^{-1/3} \rightarrow \nu_e^0 + d^{-1/3}$

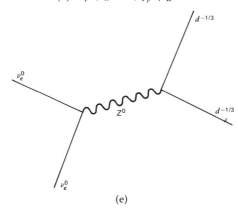

(e)

Figure 6   Elastic scattering of neutrinos and either protons or neutrons, and its interpretation in terms of quarks: (a) neutrino-proton elastic scattering; (b) neutrino-neutron elastic scattering; (c) elastic scattering of a neutrino and a $u^{+2/3}$ quark in a proton; (d) elastic scattering of a neutrino and a $d^{-1/3}$ quark; (e) Feynman diagram for the interaction given in (d); note that the reaction proceeds through the interchange of a hypothetical z^0 particle.

shown in Figure 6. The neutrino interacts with a u quark to give just a neutrino plus u quark; and similarly with a d quark. In terms of Feynman diagrams, Figure 6 shows a neutrino going into a neutrino and a d quark into a d quark with the exchange of a heavy, electrically neutral quantum called Z^0. Like the X quantum, we hope to see this quantum some day in high energy experiments. The exchange of this quantum gives rise not only to a force between a neutrino and a quark but also between the neutrinos, between neutrinos and electrons, between electrons and protons, and so forth. An understanding of this force is essential in order to complete a unified flavored dynamics. But that is probably not all. Probably, the unified flavor dynamics will require many other new forces, including

Color \ Flavors	d Flavor	u Flavor	
Red	$d_R^{-1/3}$	$u_R^{+2/3}$	
Green	$d_G^{-1/3}$	$u_G^{+2/3}$	} Gluons
Blue	$d_B^{-1/3}$	$u_B^{+2/3}$	
None	e^-	ν_e^0	

Figure 7 Several particles (quarks, e^- and ν_e^0) and their colors and flavors. Color forces are carried by colored gluons, such as a red-blue gluon or a blue-green gluon. The theory of color forces is called quantum chromodynamics. Flavor forces are carried by γ^0, $X^\pm$, and Z^0 particles. The theory of these forces is called quantum flavor dynamics.

ones of shorter range involving heavier quanta. I expect that it will be a large mathematical system.

Now let us review what we have said so far about flavor forces, color forces, and the quark and electron-like particles. In Figure 7 we have a list consisting of the electron and its friend, the electron neutrino, two different flavors of electron if you like. Then we have red, green, and blue quarks, each coming in two flavors. Notice that the electric charge varies with flavor. Among the colored quarks, red, green, and blue, there are strong color forces carried by gluons, and we think these are described by our unique mathematical theory, QCD. Connected with the flavors we have quantum flavor dynamics, a theory which is not yet completely formulated, but which attempts to describe the flavor forces. The quanta of the flavor forces are the photons which carry the electromagnetic force, the X^+ and X^- which carry the weak force, and the Z^0 which carries the brand new neutral weak force. Probably there are some other quanta, which carry additional flavor forces, as well.

I want to remind you that particles like the electron, its friend the electron neutrino, the photon, and the various intermediate quanta for the weak interactions must be detectable directly in the laboratory, whereas the quarks and gluons are believed to be forever confined inside strongly interacting nuclear particles like the neutron and proton. While these are different kettles of fish from the observational point of view, from the theoretical point of view there are strong analogies between the quarks on the one hand and the electron and the neutrino on the other, and between the gluons on the one hand and the photons and the X and Z particles on the other.

We believe that the equations of flavor dynamics are similar in many respects to those of color dynamics but do not lead to confinement. Flavor also differs from color in that the pattern of the quark flavors, the flavor pattern of the electron and its neutrino, and the pattern of the photons and the various quanta of the weak forces show the interesting property of broken symmetry. Professor Weinberg will describe how particle physicists are beginning to succeed in explaining broken symmetry, using very simple equations which have perfect symmetry. The new concept of broken symmetry involves equations that are symmetrical, but have solutions that are not. There is an infinite symmetrical set of these unsymmetrical solutions, one of which is chosen by the world. We conjecture that a theory of that type will eventually answer the question of why the world exhibits so much broken symmetry.

Today, in theoretical particle physics, we are going beyond flavor dynamics, which we have not yet completely formulated, and color dynamics, which we think we may have formulated completely, and we are looking for a unified scheme that will embrace flavor dynamics and color dynamics and some new forces all together. The equations of flavor dynamics and those equations of color dynamics are very similar, when you take into account the spontaneously broken symmetry that I just described. We hope they can be unified by a single mathematical equation of the same type for all forces except gravity. This grand design has been for a long time our most cherished dream, and we now seem to be rather close to it.

How do you search for such a theory? Well, you try a lot of possibilities. They usually fail to be self-consistent, or they fail to agree with known facts, and so you throw them away. Finally you find something that agrees with known facts and is self-consistent. This is so difficult that by the time you have done it you are likely to be right.

Some theorists are even trying to unify all of these forces with gravity in a nontrivial and geometrical fashion, the kind of thing that Einstein tried for years but failed to accomplish. Just recently, in the last few months, that program has made a lot of progress. But there is as yet no experimental support whatsoever for the ideas of general unification. So far their appeal is entirely aesthetic. Some day, we may be able to confront them with experiment. In the meantime, as a purely mathematical conjecture, they are working fantastically and surprisingly well. We had not dreamed how easy and beautiful it was to write equations that seem to embrace, in a natural unified geometrical manner, gravity and some of the other forces that we have described. All the equations tried so far have serious flaws, but they give rise to hope that the program of complete unification might be carried through in the future.

Now let me end by describing some complexity that I have glossed over. I did not list enough flavors; there are more than just two for the quarks and there are more than just two for the electron and its friend the electron neutrino. (See Figure 8.) In fact we know that there is a muon, which is like a heavy electron—about 200 times heavier, but otherwise very similar. Also we know that there is a neutral muon neutrino that is a friend of the muon the way the electron neutrino is a friend of the electron. So there are at least four flavors here, and as I. I. Rabi has asked, "Who ordered them?"

For the quarks also, there must be four flavors. Besides *d* and *u* quarks we have the strange quark. This quark is made necessary by the concept of strangeness, which we investigated some quarter of a century ago. That makes three quarks. More re-

Color \ Flavors	d Flavor	u Flavor	s Flavor	c Flavor
Red	$d_R^{-1/3}$	$u_R^{+2/3}$	$s_R^{-1/3}$	$c_R^{+2/3}$
Green	$d_G^{-1/3}$	$u_G^{+2/3}$	$s_G^{-1/3}$	$c_G^{+2/3}$
Blue	$d_B^{-1/3}$	$u_B^{+2/3}$	$s_B^{-1/3}$	$c_B^{+2/3}$

Figure 8   Flavors and colors of quarks that are demanded by the latest experimental and theoretical considerations. The flavors of the electron-like particles are: e^-, ν_e^0, μ^-, and ν_μ^0.

cently, in order to construct a unified flavor dynamics, a fourth flavor (c for charm) was definitely predicted by Glashow et al. This prediction has now been confirmed by experiment. The quark has not been observed, of course, but the fourth flavor has been found to exist inside observable particles. While this is a complication it is also a triumph for theory.

It is not clear, however, that we are at the end. There are experiments and there are good theoretical ideas which suggest to us that there may be more than four flavors of quark and there may be more than four flavors of the electron-like particles. So the skeptics (there are some in our business, as you have heard) then begin to wonder how there can be so many elementary particles. That is, as Monty Python would say, "a fair question and one that has been much on my mind." Various answers to this question are being offered. It may be that the quarks and the electron-like particles, indeed all those particles that we regard today as fundamental, are simply composite, which would explain why there are such a lot of them. Perhaps there are some simpler subquarks and subelectrons and so on, out of which the quarks and electrons we know are composed. This idea has in its favor historical precedent. It has always been the case before that when we found apparently fundamental objects they were made of something a little more fundamental. Presumably this could go on indefinitely. However, it may not go on indefinitely. The fact that it has always happened is no guarantee that it will happen now. Also, there is so far no sign that they are composite. We know of no excited states of the electron and neutrino showing higher angular momentum, which we would expect if they were made up of constituents that can revolve around one another. Since the quarks themselves are not directly visible, we have no direct means of determining whether they are composite or not. But in the course of experiments on the neutron and proton we should be able to test whether they really act like composites. So far they certainly have shown no sign of it. However, that may simply mean, as Professor Weisskopf's paper seems to say, that one has to go to immensely high energies and vast public expense in order to detect the composite character of these particles. I would like to believe, at least for the time being, that this is not the alternative that nature has chosen. I hope rather that this time we really are dealing with fundamental objects and that there is some fundamental reason why there are so many of them. Nature may have chosen some very elegant mathematical scheme that requires just these particular numbers of flavors, colors, and so on. There are mathematical systems that require definite numbers for consistency and one of them may be relevant here. If the overall unified color and flavor dynamics approach is to succeed, that must be the way things are.

Another possibility is that we are just not asking our questions in exactly the right way. Perhaps some now unknown brilliant young scientist will find a new set of questions to ask, the answers to which will clarify today's problems and make what I have been saying here obsolete.

Is Nature Simple?

by

STEVEN WEINBERG
Higgins Professor of Physics
Harvard University

We like to think that nature is fundamentally simple, that it is governed by simple general laws. This may not be true, but it seems wise to assume so, as a guide to our work. Therefore, when we look at the current state of physics, and judge how simple nature seems in the light of our current understanding, we are testing not so much whether nature itself is simple, but how close we have come to working on the level of the fundamental laws of nature.

If you take a casual look at textbooks of elementary particle physics as they have been written for the last few decades, I think you might at first be inclined to give us a grade of something like B− in simplicity. There are great varieties of kinds of particles and of forces, which were discussed in this conference by Professors Weisskopf and Gell-Mann. On the level of observable particles we find hadrons, leptons, and certainly the photon. On the level of forces, there are the familiar forces, the gravitational force and the electromagnetic force, familiar because they have long range so that we can feel them in everyday life, and the more recondite forces, the weak nuclear force and the strong nuclear force, which have such short range that we have to study them in physics laboratories. The different particles feel the forces in different ways: the leptons feel the electromagnetic and the gravitational and the weak force; the hadrons feel all these and also the strong force; and so on. There are many, many hadrons and a fair number of leptons, and the whole picture is exceedingly complicated. I think just looking at this picture superficially one would have to conclude that we have not come very far toward a simple view of nature, and therefore that we are not very close to a fundamental understanding of nature.

However, for some time now we have been receiving intimations that beneath the level of elementary particle physics, there is yet a deeper level—call it if you like the substratum—and that this substratum to elementary particle physics is enormously simpler than you could imagine from the appearance of elementary particle physics. It is this increasing understanding of the substratum underlying elementary particle physics that I want to talk about here.

One of the features of the substratum which I want to mention, but not dwell on, is the fact that its inhabitants are probably best described not in terms of particles at all, but in terms of fields. As most of you probably know, these ideas of particle and field are not absolutely logically distinct. I think most people have gotten used to the idea that light is made out of varying electromagnetic fields and yet at the same time when you study the light emitted or absorbed by a single atom you find that it does not come in a continuous wave; it is a single chunk of energy and momentum, a particle called a photon. You could just as well say that the photon is the fundamental thing and that a light wave is nothing but a coherent state consisting of enormous numbers of photons. So there is a certain amount of logical give and take between the ideas of field and of particle. Many of the fields that we think describe the substratum do, in fact, correspond to known particles: the photon corresponding to the electromagnetic field, the electron corresponding to something else called the electron field, the neutrino corresponding to something else called the neutrino field, and so on. Also there are other particles which have not yet been observed in the laboratory, but whose fields we have encountered in our theories and which we therefore are predicting will be found in the laboratories when time and money permit. These are the W particles and the Z particles that Murray Gell-Mann mentioned, and other particles as well, things that are even more esoteric, so-called Higgs bosons and pseudo-Goldstone bosons, and so on, about which we argue endlessly.

If that were all, one might say that the two notions of field and particle were equivalent. However, we seem to be coming to a position in physics where in fact many of the fields, in terms of which the laws of physics at the level of the substratum are written, may not correspond to observable particles at all. This is particularly true of the quarks and the gluons that Gell-Mann described in the previous paper. These are the fields in terms of which the laws of the strong interactions must be written. We think of the quark field interacting with the gluon field, and therefore with itself. But the particles which are the quanta of these fields, although for some approximate purposes they can be dreamed of, will, I would guess, never be isolated in the laboratory as free particles. In fact, most of the particles that we see in the world, most of the particles which we are familiar with, such as the protons and the neutrons which make up all but a tiny fraction of our own mass, do not correspond to elementary fields. The laws of nature are not written with the proton field and the neutron field as ingredients, much less the Σ^- field or one of some other conceivable fields associated with some other recondite particles. A great measure of simplicity is achieved by replacing the description of nature in terms of enormous variety of types of particles (so many in fact, that most of us carry around a little booklet published by the Lawrence Radiation Laboratory listing particles that have been discovered from month to month) with a description in terms of fields.

The other feature of the substratum, which is the one I wanted more specifically to talk about, has to do with symmetry. The substratum underlying elementary particle physics exhibits a tremendously high degree of symmetry, much higher than you would think from the ordinary experience of elementary particle physicists. Let me say a few words about symmetry, which is central to everything else I am going to say. A symmetry is always just a statement that something appears to be the same no matter how you look at it, or in a more limited way, that there are certain different ways you can look at a thing and it will appear the same. For instance, the sanctuary light hanging above my head as I talk has a certain symmetry—I counted nine flanges and that means that I can stand here and it has a certain appearance to me. I can walk over here, walking 40 degrees around it (since 9 goes into 360 degrees 40 times), and it will appear the same, and so on. And to you over there, at an angle of 80 degrees from the direction I am facing, undoubtedly it appears the same to you as it does to me. The group of symmetries of the sanctuary light then consists of rotations by any multiple of 40 degrees.

Now the example of the sanctuary light is slightly misleading, because what I am talking about here are not the symmetries of things, but the symmetries of the laws of nature. And when I say that the laws of nature possess a certain symmetry I mean that the laws of nature allow a person to change his frame of reference without changing the form of the laws of nature. The most familiar examples of such symmetries are, like the symmetry of the sanctuary light, space-time symmetries. For example, the laws of nature allow you to rotate your laboratory frame of reference, allow you to orient your laboratory without paying any attention to advice from geomancers, in any direction you like. And the laws of nature will remain the same—not only for rotations by 40 degrees, but rotation by any angle you like. And if you study the laws of nature in a rotated laboratory you will find the same laws of nature. We call that rotational symmetry, or rotational invariance. Another familiar symmetry (which the sanctuary light does not have) is translation symmetry. The laws of nature appear the same wherever you put the laboratory. I can put it here, or I can put it in Stanford or Cambridge, and find the same laws of nature. And finally the laws of nature appear the same when you translate in time. They are the same laws now as they will be next year. Another way of saying this is that I could reset my clock and repeat all my measurements and the laws of nature would appear the same.

From these symmetries we derive consequences. From these three in particular we derive conservation laws. The law of rotational symmetry mathematically implies (in a way that I can not go into here) the conservation of angular momentum, of spin. The law of invariance under translations implies the law of conservation of ordinary momentum, mass times velocity. And the law of invariance to the date at which you set your clock implies the conservation of energy.

Notice what is happening here. We do not know the laws of nature. These symmetries are statements about the laws of nature which are presumably true even despite our great ignorance about what those laws are. From those statements we derive consequences and those consequences, at least the ones I have mentioned so far, are verified as far as we know to an unlimited precision.

Now in addition to the symmetries of space and time, there are other symmetries which you can imagine, some of which have been discovered to be actually true. For example, in nuclear physics it has been known since the 1930s that there is a high degree of symmetry between the two types of particles which form the constituents of atomic nuclei, the proton and the neutron. You can take a nucleus (this is approximate) and replace all protons by neutrons and vice versa and have another nucleus which appears for most purposes the same. This works particularly well in some nuclei; for example the heaviest isotope of hydrogen—^{3}H—consists of one proton and two neutrons. There is another isotope—^{3}He—which is a light isotope of helium consisting of two protons and one neutron. In fact, all you have done in going from one to the other is to exchange a proton with a neutron, and in consequence they have very nearly the same mass.

We can state this as a symmetry. The laws of nature, if we want to write them in terms of protons and neutrons (which I said earlier we do not), would retain their form if we replaced every proton in the universe by a neutron and vice versa; or in terms of fields, if everywhere in our equations when a proton field appeared we replaced it by the neutron field and vice versa. (You have to do other things too to other particles, and I am oversimplifying, but it is close enough.)

In fact, because of the wonders of quantum mechanics you can do even more than that. You can contemplate symmetries in which you do not just change a proton into a neutron but you change a proton into a state which has a 30% chance of being a neutron and a 70% chance of being a proton. Quantum mechanics allows you to think of states which are not definitely one thing or another. You can, if you like, have a state which has a 31% probability of being a neutron and a 69% probability of being a proton, or what you will. In fact, mathematically, the description of this symmetry is in terms of continuous rotations. These are the "internal" symmetries. (It is not that we think that the proton and neutron really have an internal structure. It is rather just the name we give to this particular class of symmetries.)

Now, there are a number of these internal symmetries and

every one of them again generates a conservation law. The symmetry between the proton and the neutron generates a fairly unfamiliar conservation law, which is called isotopic spin conservation. (But notice the analogy between spin in this sense and spin in the ordinary sense of angular momentum.) The conservation of charge is one of these conservation laws that comes from a similar internal symmetry principle. And there are others. The conservation of "color" is another one.

Now that is not all. I have talked so far about space-time symmetries and about internal symmetries. The two ideas get mixed up in modern physics in a very fruitful way. You can imagine changing your frame of reference in a way that depends on where and when you are. When I have talked so far about changing a proton into a neutron I have said the laws of nature retain their form (approximately) when every proton in the universe is changed into a neutron and every neutron in the universe is changed into a proton. But suppose you change protons into neutrons in different ways and in different places—here and now I leave them alone; last week in Stanford, I changed protons into 70% protons and 30% neutrons; and next week in Cambridge I do something else. This can also be contemplated as a possible symmetry of nature. Such symmetries have been considered for a long time. They are usually known in the trade as gauge symmetries. (The word "gauge" comes from the usual English word for "measure," as in gauges of nylon stockings or gauges of model railroads. But the connection is obscure at best and I will not go into it; it is just what they are called.) These gauge symmetries, unlike the space-time symmetries and the internal symmetries that I have mentioned before, do not imply conservation laws. Instead they have a much more remarkable consequence. The gauge symmetries imply the existence of fields and govern the interactions of those fields.

For example, the gauge symmetry which we find in the theory of electromagnetism in fact requires the existence of the electromagnetic field. It requires the existence of the photon and it governs the interactions of photons. It is hard for me to explain, without mathematics, how this could be. The only analogy that I have been able to find, which I think I can explain at all in nonmathematical language, is with gravitation. So let me try it, and if I do not quite succeed in making this point clear it is my fault and not yours.

I said that the difference between a gauge symmetry and an ordinary internal symmetry is that an ordinary internal symmetry says that the form of the laws of nature do not change if you do the same thing everywhere and at all times, if for instance you change neutrons into protons and vice versa, while for a gauge symmetry the equations of motion do not change even if you do different things at different places and times. Now consider, as the symmetry you want to think about, not these symmetries involving protons and neutrons, but the symmetry of rotational invariance. That is the symmetry which says the laws of nature look the same in whatever direction your laboratory is oriented. Up till now when I have talked about rotational symmetry I have imagined rotating 40 degrees, or 70 degrees, or what you will, but rotating everyone's laboratory all over the universe in the same way. But suppose you do not do that. Could rotational invariance be a gauge symmetry? Instead of talking about different places let us just talk about different times—which works the same way. Suppose I considered rotating the laboratory by a certain angle, but an angle depending on time. In other words not just rotating the laboratory and then repeating my experiments but rotating the laboratory continually, so that I keep spinning around. Will the laws of nature appear the same in a spinning laboratory? At first glance, the laws of nature do not appear the same: the simplest manifestation is that if a person spins on his vertical axis his arms are drawn up by centrifugal force. It thus appears that the laws of nature cannot possibly respect rotational invariance in the gauge sense, that is it cannot be invariant under rotations depending on position in space and time. But that is not really the end of the story, because there is an alternative explanation for what produces the force on my arms. That is, instead of supposing that I am spinning I can take an extraordinarily self-centered point of view and imagine that the whole universe is spinning in the opposite direction. And this enormous motion of matter and energy all around me sets up a gravitational field which pulls my arms up. Within the mathematical framework of general relativity, one can show that this is a quite tenable interpretation of centrifugal force. So you see that because of the existence of the gravitational field as an agent in nature, one is capable of saying that the laws of nature do not depend on your frame of reference even if you go into a spinning labora-tory. The symmetry is only possible because of the force of gravity, because of the existence of the gravitational field. We can now turn that around and say that the symmetry *requires* the existence of the gravitational field. In the same way there are gauge symmetries that require the existence of the electro-magnetic field, and there are gauge symmetries that require the existence of the color gluon field that was described here by Gell-Mann.

Now if this were the end of the story we would still not have gone very far. Looking at the elementary particle physics that we see in the textbooks you do not see symmetries which are all that powerful. On the surface, there does not seem to be any gauge symmetry connected with the weak interactions, for example. For another thing the symmetries we do see are not even quite true. The proton and the neutron are not exactly the same; the neutron is 1.3 million electron volts heavier than the proton. So the picture of a universe ruled by symmetry is only a hope which seems to be denied as soon as it is offered.

The reason for suspecting that the substratum underlying elementary particle physics, the field theoretic substratum, is governed by symmetry, has to do with a phenomenon which has become known by the rubric "broken symmetry." This is the idea that it is possible for the fundamental field equations to have a symmetrical form and yet for the solutions of these field equations not to share that symmetry.

I will try to explain this by giving an example, based again on the sanctuary light over my head. The laws governing the force between atoms of copper and tin (I presume that the light is made out of bronze) are translationally invariant. That is, every pair of copper or tin atoms if put a certain distance apart will attract each other or repel each other exactly the same way, wherever that pair is put in the universe. The force depends on the relative separation between them, but the force does not depend on where that pair is. You can think of any particular bronze object as a solution to equations implied by the translationally invariant force laws. It is a particular set of copper and tin atoms whose positions obey the force laws, and solve the equations of statics that our knowledge of the force between copper and tin atoms provides us. But the solution manifestly breaks translational symmetry, that light is here; it is not over there. This is the most commonplace thing in the world: simple equations often have complicated solutions.

You see this again in the flow of water past a sphere. If you have a perfectly smooth flow of water past a perfectly smooth sphere then below a certain speed the water will flow smoothly and respect the evident symmetry of the problem, which is rotational symmetry around the direction of the flow of the water. Since both the sphere and the water have the same symmetry, it is a symmetry of the whole problem and the flow will respect it. If you get up to a high enough velocity, then suddenly the flow becomes turbulent, eddies break away and the symmetry is evidently lost. In this latter case, the solution of the equations of hydrodynamics in a completely symmetrical situation do not exhibit the symmetry of the underlying equation.

How do we know about these broken symmetries? This is not so easy. In fact it has not been easy historically. The great symmetry of translation invariance which says that the laws of nature are the same wherever you put your laboratory certainly was not evident throughout most of the history of science.

It is often very hard to infer the existence of a symmetry describing the underlying laws if that symmetry is not realized in the solution that you actually have to deal with. In particle physics the manifestations of broken symmetry depend on the details of which symmetry is being broken. This is a complicated story, which was worked out during the 1960s with a good deal of difficulty, and I will only mention one aspect of it here.

If a gauge symmetry (remember that is a symmetry in which the transformation can depend on space and time) is broken, then the field which is necessitated by the gauge symmetry—in the sense that the electromagnetic field is necessitated by one gauge symmetry, and the color gluon field by another, and the gravitational field by another—the field which is necessitated by that gauge symmetry acquires a mass. The photon has zero mass, because that gauge symmetry is not broken. The color gluon presumably has zero mass because color symmetry is not broken. But the fields which are necessitated by those gauge symmetries which are broken acquire masses. This then is to be the natural explanation of the weak interactions. The weak interactions are transmitted by the exchange of particles, which are the quanta of a field which is related by an exact symmetry to the other gauge fields, to the photon gauge field in particular. But that symmetry is broken and therefore we do not see it in ordinary life. The weak interaction appears very much weaker than the electromagnetic interaction simply because the particles that transmit it, the W particle and the Z particle, are much heavier than the photon. Their masses are figured out from the fact that at ordinary energies the effects of the weak interaction are extremely small, but the calculation actually uses the idea that there is an underlying broken symmetry between the weak and electromagnetic interactions. This line of reasoning then allows one to predict the properties of a whole new class of weak interactions, the "neutral currents," which are produced by exchange of Z particles. To test these ideas, an experimental search for the neutral currents was carried out in 1973, and they were found, with just about the expected properties.

I want to emphasize that I have been talking about the breaking of internal symmetries. As far as we know the space-time symmetries are not broken. They are exact. There may however be larger space-time symmetries which are broken. It may be that the whole geometry of space and time that we are familiar with just deals with that little subspace of some larger space that is governed by symmetries that are not spontaneously broken. But about this it is very difficult to speculate, and there have not really been any fruitful speculations.

One of the intriguing features of the idea of broken symmetry is that it is possible in a sense to restore the symmetry under suitable physical conditions. You can go to very high energy. You see that the symmetry between the weak and the electromagnetic interactions (which I repeat is a symmetry on the level of the field equations governing the substratum) is broken and that breaking is manifested in the appearance of a mass, of the order of about 80 gigaelectron volts—80 times the mass of the proton—for the particles, the W particles and the Z particles, that are the siblings of the photon. If you go to an energy which is sufficiently high compared to that, build an accelerator which can go to an energy sufficiently high compared to 80 gigaelectron volts, and then do experiments, the symmetry will be obvious. Those experiments, of course, are hard. But the symmetry is restored in a sense, in an approximate sense, with an increasingly accurate approximation, by going to high enough energy.

This point also finds an analogy in the sanctuary light above my head. I can restore its symmetries. Remember I said (and now I am going back from the symmetries of the laws of nature to a more concrete example of the symmetries of an actual physical object) that the light has only a symmetry under rotations of 40 degrees. There are not flanges in every possible direction—there are only nine flanges and they are 40 degrees apart. (Just to remind you of the points I made before, the laws of nature and in particular the laws governing the force between copper and tin atoms are invariant under any possible rotation. They do not care how the laboratory or the sanctuary light is oriented. So that object represents a broken symmetry. It represents the breaking of rotational symmetry down to a smaller group of symmetries, rotations only by multiples of 40 degrees.) Well you can restore the symmetry of the sanctuary light by heating it. If you heat it enough, at a certain point it will evaporate—I guess it melts first but let us skip that. It evaporates, and the vapor will fill the room. After it settles down and all the eddies and turbulence die out, the vapor will appear the same wherever you are in the room as long as you do not get too close to the walls, and in whatever direction you look. In other words, we can restore the symmetries of the laws of nature that are manifested by an object—remember, an object represents breaking of the symmetries of the laws that govern the constituents of the object—by causing a phase transition, an evaporation.

The same has recently been realized to be true in elementary particle physics. If you heat matter to a temperature of the order 300 gigaelectron volts, there will be a phase transition. In fact it is what is called a first order phase transition, in which suddenly, not gradually, the symmetry is regained. At this critical temperature the symmetry between the weak interactions and the electromagnetic interactions would be restored.

Now we can not really do that in the laboratory. We can make energies of the order of 300 gigaelectron volts in laboratories today, but we can not really get temperatures that high, because the idea of temperature requires that things have really settled down to a kind of equilibrium. We only get these energies in the laboratory in brief instants.

However, we believe that there was a time in the history of

the universe when the temperature of the whole universe was that high. This time can be estimated quite reliably. (This is one of the things that according to quantum chromodynamics, we really think we now know how to calculate.) It was about a million-millionth of a second, after what we call the beginning. At all times during the first million-millionth of a second (10^{-12} second) the temperature was above 300 gigaelectron volts and the symmetry between the weak and electromagnetic interactions was manifest. The W particle and the Z particle had zero mass and were just like the photon. They were all siblings together, quite indistinguishable.

The other symmetries that we so far only dream of could presumably be restored at even higher temperatures. There is a symmetry which some of us have begun to think about, a symmetry which relates the weak and the electromagnetic interactions to the strong interactions. Almost certainly, in order for this symmetry to be restored we must go to enormously higher energies and enormously higher temperatures—it is not clear how high. It may well be that the energies are at the level where gravitation is beginning to be important.

You see, gravity is produced not just by mass, but by energy. The sun has a gravitational field which is slightly stronger than it would be if the sun were cold because the energy in the heat of the sun also adds a little bit to its gravitational field. The earth is going around the sun a little bit faster than it would if the sun were not hot. If you make objects sufficiently energetic, then it is their energy rather than their mass that provides the source of the gravitational field, and eventually you can get up to an energy so high that gravity becomes as strong as any of the other forces. Some of the calculations that have been done seem to indicate that the point at which the symmetries among the strong, weak, and electromagnetic interactions become restored, is at those extremely high energies.

There is another point, which I mentioned earlier, about the ordinary world of physics as we see it in elementary particle physics textbooks. The symmetries in addition to being apparently of limited scope are apparently approximate. I am now not talking about symmetries like that between the weak and the electromagnetic interactions, which are not apparent, but as far as we know are exact. I am referring here to symmetries (like the isotopic spin symmetry between the proton and neutron) which are quite evident, and have been in the textbooks of elementary physics for many years, but are clearly not exact. My personal feeling is that as a consequence of the gauge theories, all these approximate symmetries can be, and in fact are being, understood as dynamical accidents. They are all in a sense fictitious symmetries. I think this applies, not only to the symmetry between neutrons and protons, but even to symmetries like the one between right and left.

I did not mention this one before, when I was talking about space-time symmetries, but you know that in addition to making rotations you can also interchange right and left, and ask "are the laws of nature the same?" It was a traumatic discovery in physics in 1956 that they are not the same, exactly, but they are nearly the same. That is probably the most spectacular example of an approximate symmetry. It may be quantum chromodynamics that explains why the symmetry between left and right is as good as it is. A quark can either be spinning to the right or spinning to the left. A gluon interacting with the quark only sees the color of the quark, and the color of a left-handed and right-handed quark are the same—red, green, or blue. So the gluon interacting with the quark (because of the stringencies imposed on that interaction by the gauge symmetry) cannot tell whether it is a right- or left-handed quark. In that way, I think, the approximate conservation of parity can be understood. Parity is not conserved by the weak interactions because they work in a different way. The flavors, as far as the weak interactions see them, of left- and right-handed quarks, are quite different and those flavors' interactions with the W and Z particle do distinguish right and left.

It is going to take a while to see how all this turns out. It is probably too early to try to think of drawing any philosophical lesson from discoveries in physics. But there is one lesson that I am tempted to draw. It is just the old familiar lesson, that things are often simpler than they seem. The laws of nature give a fundamental role to certain entities. We are not really sure what they are, but at the present level of understanding they seem to be the elementary quantum fields. They are highly simple because they are governed by symmetries. These are not objects with which we are familiar. In fact, our ordinary intuitive notions of space and time, causation, composition, substance, and so on really lose their meaning on that scale. But it is just at that scale, at the level of the quantum fields, that we are beginning to find a certain satisfying simplicity.

An Astronomer's View
of the Evolution of Man

by

SIR FRED HOYLE

Plumian Professor in Astronomy and Experimental
  Philosophy
Cambridge University

To a young child an hour seems a long time, a week seems an age, and a year an eternity. Now modern man sees time somewhat similarly. Four years, from one election to the next, seems a long time to our governments. A human lifetime seems an age to each individual and a few centuries seems an eternity. Yet just as a young child's experience of life is very short, so the total time span of civilized man is very short. There have only been 200 generations of human civilization, whereas there were upwards of 10,000 generations of human prehistory. And if you count 30 years to a generation, then life has existed on the earth for more than a 100 million generations.

There is another characteristic which modern civilized man shares with a young child. We have the impression of a future in which almost anything seems possible. But as a child reaches the age of about eight years it begins at last to see that not everything is really possible. It begins to distinguish between what might be and what can never be. Civilized man has not yet quite reached a comparable age of discrimination, although the first seeds of doubt appear to be forming right now—in this second half of the twentieth century. And indeed in my talk tonight I am going to discuss a major constraint on the future, a constraint which in its very nature appears inescapable.

It may seem strange that one can assert a limitation, not just for the present moment, but for all time, essentially however far we wish to go forward into the future. But the constraint comes from the materials in the world around us, materials taken on a cosmic scale. The relative abundances of the elements as they exist between the sun and stars, in the gases between the stars, and probably also in galaxies other than our own, are shown in Figure 1. True, the different stars we see in the sky vary a little one from another in their chemical compositions, but not by factors that would make much of a change when they were displayed on the logarithmic scale shown in this figure. Notice that each unit on the left-hand scale corresponds to a factor 10 in relative abundance, so that atoms of hydrogen and helium, the commonest elements, are about 1000 billion times more abundant in nature than the least abundant elements. The least abundant elements are those of largest atomic weight. That is to say, the least abundant elements are those whose atoms have nuclei which contain the largest numbers of protons and neutrons.

Let me give you a few examples of the economic importance of these abundances. The commonest and least expensive metals, magnesium, aluminum, and iron, have rather high abundances. Zinc and copper, being less abundant, are more valuable. Tin is still less abundant, and still more valuable. The

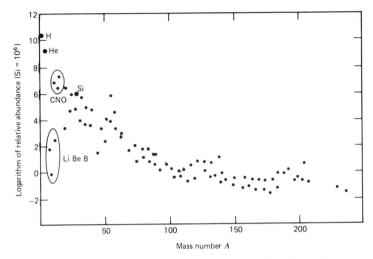

Figure 1   Relative abundances of the elements as a function of their mass number.

precious metals, silver and gold, platinum, palladium, rhodium, and iridium, are all cosmically rare elements. Of course, the relation between economic value and cosmic abundance is not entirely strict. There is also some dependence of economic value on the physical properties of the elements themselves and there is a dependence on the processes in the earth which have led to the formation of deposits in the surface rocks of the Earth. Some elements tend to concentrate into comparatively rich deposits, others remain more diffusely distributed and are therefore economically more expensive to extract from the rocks.

It is interesting and important that life itself is based on the commonest elements, particularly hydrogen, carbon, nitrogen, and oxygen. Of all substances, water is perhaps the most essential to life, and the molecule of water, H_2O, is composed of this cosmically most abundant element (hydrogen) and the third most abundant element (oxygen). Carbon and nitrogen

Figure 2   Photograph of a cosmic gas cloud.

form the backbone of living material, and these are the fourth and fifth commonest elements, respectively. Other important elements, calcium in our bones, iron in the blood, magnesium in the green chlorophyll of plants, sulfur, and phosphorus, are all elements of rather high abundance, much higher than the elements of large atomic weight. The rarity of the elements of large atomic weight prevents them from playing important roles in the structure of life. Life is built from the simplest molecules containing the commonest elements, suggesting that on a cosmic scale life may be very abundant indeed.

Not only are the basic life-forming elements found in cosmic gas clouds, like the one shown in Figure 2, but to the surprise of most astronomers the life-forming elements are themselves associated into organic molecules in the denser part of these clouds. At the last count, which I made, 36 different organic molecules had been discovered by the new techniques of "millimeters-waves" astronomy.

The discovery of organic molecules within dense interstellar clouds has added impetus to the idea that life may have preceded the formation of our own solar system. The Sun and planets have ages in the range of $4.6–4.8 \times 10^9$ years, whereas our galaxy, the whole system of the Milky Way, has an age of about 12×10^9 years. The thought is that life here on Earth could have been derived from biologically important molecules which existed already in this gas cloud from which our solar system formed. If so, just the same biological conditions would apply for many other newly forming stars, with the implication that life in our own system may be related to life in other systems. According to this point of view, life would not be required to originate independently in each stellar system.

The dense interstellar clouds also contain myriads of fine grains, with typical diameters of about $\frac{1}{20}$ of a micron. When such a cloud happens to be projected against a bright background we see it in a dark outline, as with the spectacular example shown in Figure 3. Organic molecules that stick to the grains can join together, and it is interesting that the joining of two such molecules, formaldehyde (HCOOH) and methanimine (CH_2NH), both of which are known to exist in large quantities, yields the simplest and the commonest of the amino acids, glycine (NH_2CH_2COOH). Temperatures within the clouds take values at which amino acids have been found to polymerize into chains, the resulting polymers having a sticky quality that would cause grains to adhere together. It is of considerable interest that the perimeter of a single grain is comparable to the scale of a biological gene, while the clumps of grains caused by their sticking together have a scale comparable to the simplest biological cells, a scale of a few microns. Moreover, the matrix of solid grains in which the polymers thus come to be

Figure 3   Photograph of a cosmic dust cloud in front of a highly illuminated background.

embedded provides an important shield against radiation which would otherwise break up the comparatively fragile organic structures.

These considerations make the suggestion of an interstellar origin for life much more plausible than it seemed only a few years ago. Indeed it has become significantly less difficult to see how the simpler organic molecules can become associated together within dense interstellar clouds than it is to see how this might have happened here on the Earth. And in further support of this view, both amino acids and occlusions looking very much like interstellar grain clumps have actually been found in a certain class of meteorites, the carbonaceous chondrites as they are called. So the interstellar theory begins to look like the frontrunner rather than merely a quaint speculation.

But to come back to the abundances of Figure 1, you may wonder why the abundances are so universal. What made abundances with this particular distribution, high for most elements of small atomic weight and low for elements of large atomic weight? The answer to this important question lies in two parts. The first part concerns a state of the whole universe, which occurred about 15 billion years ago, when all material was highly compressed and very hot. The second part has to do with detailed physical processes that take place within the stars. Let us consider then these two parts in turn.

The hot dense state of the universe to which I have just referred is often called the "origin" of the universe, although whether it was really the origin of all things is a matter for debate. What is not a matter of debate is that the very high temperatures which existed would prevent protons and neutrons from associating together, as they do in the nuclei of atoms under normal circumstances. Any nucleus built from several protons and neutrons would have been evaporated immediately by the intense radiation field. The high temperature of the radiation field also has the effect of speeding up interchange relations between protons and neutrons, so that a balance is set up between these two kinds of particles. Then as time goes on the universe evolves toward lower temperatures, and eventually the neutrons are able to join with protons, producing helium. There is a considerable excess of protons remaining, however, and this excess yields hydrogen, which thus comes to be the most abundant of the elements.

Although hydrogen and helium were the main elements to emerge from this "origin" of the universe, small traces of other elements may also have been present. But for the main concentrations of the other elements we must now turn to the stars, to the second part I referred to a moment ago. To sum up the first part, the effect of the "origin" of the universe, some 15 billion years ago, has been to immerse us in a predominantly hydrogen-helium environment. This has provided, as we shall see in a moment, for the energy of the stars, and for a complex of processes to occur within stars that have had the effect of synthesizing the rich variety of atoms we find in our everyday world.

Stars are known to form by condensation in gas clouds like that of Figure 2. Gravitation pulls the materials of protostars together, compressing material and causing a rising temperature within it. Eventually the rising temperature leads to energy production by nuclear reaction. What are these reactions?

Historically, this question was a hard one to answer, mainly because of a wrong nineteenth century belief that one kind of atom cannot be changed into another kind of atom. Scientists of the nineteenth century thought every separate kind of atom was indestructible in itself.

The discovery of the natural radioactivity of the heaviest elements around the beginning of the present century showed that atoms are not indestructible, however, and so gave the beginning of the right idea for explaining the source of the energy emitted by the Sun and by other stars. Yet this radioactivity of the heaviest elements, for example, uranium and thorium, could not possibly be responsible for this energy, just because the abundances of the heaviest elements are so small—there is far too little of them. This difficulty caused J. Perrin in 1919 to suggest that if four atoms of hydrogen could be converted into one atom of helium, energy on an adequate scale could be released, precisely because hydrogen is so very abundant in the world.

This idea has turned out to be correct. Indeed it *had* to be correct, for there is no other way, consistent with the abundances of the elements, to explain the very large energy supply which the stars clearly possess. Unlike the situation in 1919, today we understand how hydrogen can be changed to helium. There are two main processes for this, which go by the names of the *proton-proton chain* and the *carbon-nitrogen cycle*.

The generation of energy inside stars has an inevitable quality about it. Protostars forming within gas clouds like that of Figure 2 are drawn together by gravitation, until compression raises their internal temperatures high enough for nuclear reactions to change hydrogen to helium at a rate that is sufficient for an energy balance to be set up, a balance between what is produced by the nuclear reactions and what is being lost through radiation emitted into space from the outer surface—a star, rather than a protostar, is then said to be formed. Depending on the quantity of material in the star, the mass of the star as we usually say, the star occupies one position or another in the line shown in Figure 4, a line which astronomers call the main sequence. In this figure we have the rate of energy loss from the outer surfaces (in terms of the energy emission from

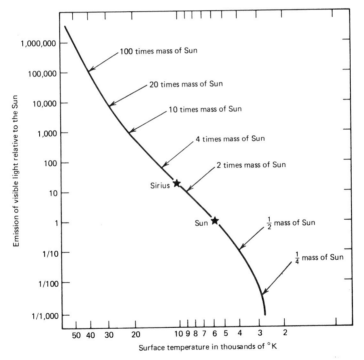

Figure 4 Luminosity of stars as a function of their surface temperature. The solid line indicates the main sequence.

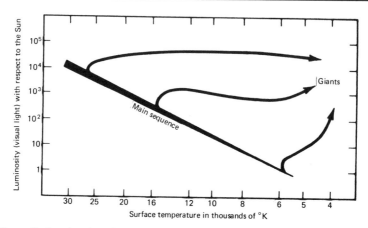

Figure 5   Luminosity of stars as a function of their surface temperature. This diagram shows schematically how the luminosity and surface temperature change over the lifetime of a typical star.

the Sun as the unit) plotted logarithmically on the left-hand scale, and the temperature of the outer surface (plotted from right to left) on the bottom scale.

Although a very great deal of energy, enormous compared to human experience, can be got from this hydrogen-to-helium conversion, there is an evident limit to it, imposed by the availability of the hydrogen. Sooner or later the hydrogen gets used up, and particularly is this the case for stars of large mass, which are the highly luminous ones lying on the upper left-hand part of the main sequence. As hydrogen becomes exhausted in the central regions of stars an evolution of this kind shown in Figure 5 takes place. Depending on where it starts, that is to say on its mass, a star moves in our diagram toward a region at left center of this diagram. Stars in this part of the diagram are large in size, of low surface temperature with a red color, and are known as Giants. By the time this happens a star is effectively in its death throes. The detailed behavior of an individual star is a good deal more complex in its detail than the

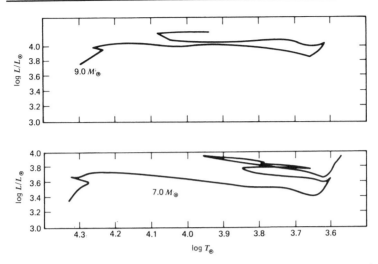

Figure 6   Luminosity-surface temperature history for a star of seven times the solar mass and for a star of nine times the solar mass. The quantity $M_\odot$ refers to the mass of the sun; $L/L_\odot$ is the ratio of the luminosity of the star to the luminosity of the sun; and $T_\odot$ is the temperature of the sun.

schematic lines of Figure 5. Thus Figure 6 gives details, one track for a star with seven times as much mass in it as the Sun has and the other star with nine times as much material.

The cause of this complex behavior is that with hydrogen becoming exhausted, and with energy from hydrogen to helium conversion becoming less and less available, the stars are seeking to derive energy from other nuclear processes, of which the first to occur produces carbon and oxygen from helium; then the carbon, nitrogen, and oxygen are themselves involved as nuclear fuels, producing a range of well-known elements— sodium, magnesium, aluminum, and silicon to name the more important ones. Finally, the silicon so produced itself serves as a fuel, yielding essentially all the elements with atomic weights between 30 and 60. These successive steps are evidently inter-combined, with the product of one stage serving as the fuel for the next stage. So we have a set of successive steps in the chemical evolution of material.

You may wonder what happens beyond the last stage, the one in which silicon serves as a fuel, with elements up to atomic weight about 60 as the product. Is any further evolution, perhaps to still larger values of the atomic weight, still possible? The answer to this question is no. Once silicon has been consumed, we reach a situation in which the original material, the hydrogen-helium environment, has been entirely converted into nuclear ashes. No further energy can then be obtained. Indeed, energy would have to be *supplied* to produce atoms with nuclei of still greater atomic weight. Only in very rare situations does this happen. Because of the rarity of these situations, the atoms of greater weight have only the very low abundances shown in Figure 1. And because of their rarity these situations are of no general help to the stars in their avid search for energy.

So the stars in their evolution are set upon a "no-win" course. However complicated its behavior, there is no gainsaying the fact that the energy supply of a star is limited, that it tends to exhaust its energy at an increasing rate, and that the energy production from nuclear reactions must eventually fail to make good the losses from radiation into space. This raises the problem of the final fate of a star, of its ultimate graveyard.

Before I come to this problem, let me show that these issues are not a matter of theory only. Figure 7 is a plot of the positions in our diagram of many stars which you can actually see for yourself with the naked eye. The crosses in the diagram are for the nearest stars, and the circles are for the brightest stars in the sky—many of them well known, such as Rigel, Vega, and Arcturus. It is a remarkable fact that most of the nearest stars are of small mass like the Sun. They tend to be intrinsically faint, and very few of them appear among the brightest stars. The brightest stars are undergoing precisely the kind of evolution we have just been discussing. Many of the brightest stars are already in their death throes.

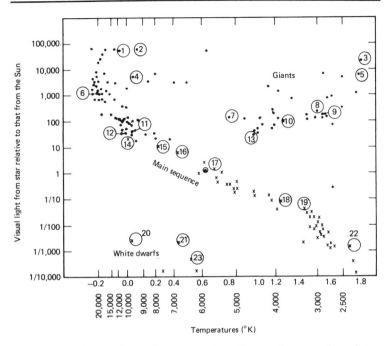

Figure 7   The luminosity-surface temperature diagram for a number of stars which are visible to the naked eye.

The problem of the ultimate fate of the stars, especially of the stars of large mass for which the question is particularly asked, is a complex and fascinating one to astronomers. The road in some cases may lead to those mysterious objects known as *black holes*. In other cases, the road leads to explosions which disintegrate most, if not the whole, of the stars. The violence of these explosions, *supernovae* as they are called, in energy terms is equivalent to 10^{26} man-made hydrogen bombs all exploding together. Shown schematically, we have the situation of Figure 8. From our present point of view, these explosions serve to broadcast far and wide in space the complex atoms which are synthesized within the stars—the carbon, nitrogen, oxygen, magnesium, silicon, sulfur, calcium, iron, and many other elements of our everyday world.

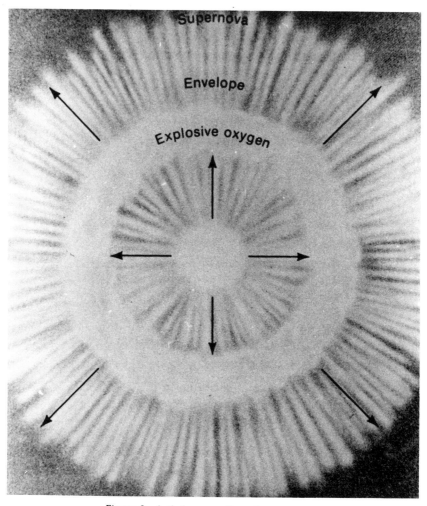

Figure 8   Artist's conception of a supernova.

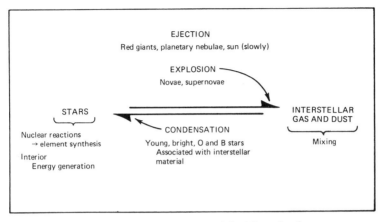

Figure 9   A schematic diagram of the life cycle of stars.

This broadcasting of complex atoms through space sets up the cyclic relation of Figure 9. Through Figure 9, we are now in a position to understand how it comes about that, although nuclear processes within the Sun are limited to only hydrogen-to-helium conversion, with no solar production of elements like carbon, oxygen, magnesium, silicon, iron, and so on, our solar system nevertheless contains all these elements. The explanation is that all these other elements were produced in stars which completed their evolution before the Sun and our planetary system were formed. Our galaxy (the Milky Way) is about 12 × 10⁹ years old, nearly three times the age of our solar system. The preceding generations of stars broadcast their synthesized materials throughout our galaxy, through an arm of the loop of Figure 9. The synthesized materials were therefore present in the gas clouds from which the Sun and planets were formed.

I began this lecture by assuming that whatever uncertainties there may be about the future of mankind, one thing is certain, namely the chemical environment in which we must operate. And now you see why this is so. Our environment, determined by the abundances of the elements, has been produced by the grand scale march of events. The scene for it was set in the first place by the "origin" of the universe. To this we have added the wide sweep of the evolution of the stars. It is useless to think that we can escape from our chemical environment by going some other place, because all places of which we can conceive have been similarly affected.

It is true that the Earth itself is not an entirely typical place. There is no enormous preponderance of hydrogen and helium on the Earth, and taken as a whole the Earth is considerably deficient in other biologically important elements, in particular, carbon and nitrogen. These deficiencies mean that the process which led to the formation of the Earth somehow contrived to produce a selection against the lightest of the elements—more precisely, against volatile elements existing as gases rather than as solids. But these deficiencies are largely made good by the circumstance that we humans live at the surface of the Earth, in the *biosphere* as it is usually called, where the Earth's hydrogen, carbon, and oxygen are concentrated. The deficiencies in our everyday world are not therefore by any means as great as they are for Earth as a whole. In effect, the biosphere returns us to an essentially cosmic environment, the environment in which life may very well have begun.

I come now to the last part of my lecture. Given then that our chemical environment cannot be much changed, what does this tell us about the long-term future of our species? A big question certainly, but one about which a surprising amount can be said. Our chemical environment imposes constraints which can be summarized in three rules:

1.   Any material which Man uses in large quantity must be made up of elements of large cosmic abundance.

2.   Elements present in lower abundances can be important economically, provided their use is confined to high technology, where small quantities of a material with special properties can play a significant role (as on surfaces or at contact points).

3.   Elements present in lower abundances can also be im-

portant economically if they are sources of nuclear power, because nuclear processes are on the order of a million times more energetic than chemical processes, which multiplies a millionfold the effectiveness of nuclear materials like uranium and thorium.

Let us see what happens when the first of these rules is contravened. Imagine a watery solution of mineral substances within the Earth's crust, squeezed at high pressure through interstices in a solid, but hot, rocky matrix. Conditions for the precipitation of various materials can be exceedingly specific, and conditions happen to be just right at a certain place for the precipitation of a certain rare element, gold. So a vein of precious metal comes to be formed. Earth movements over hundreds of millions of years then turn over the crustal rocks, causing the vein of gold to come close to the surface. Weathering at the surface now proceeds on a much shorter time scale, eventually washing some of the gold into the bed of a river, where it is noticed by an enterprising prospector. A gold rush ensues. Towns and populations grow explosively closely by that river. The boom is short-lived, however, although nobody actually in it believes the prosperity will ever end. But the boom must come to an end, because the amount of gold was never large. So towns and populations must soon wither away, unless the economic capital provided by the gold has been wisely used to establish a different, and more stable, activity for the people.

The oil boom of the present century has the quality of a gold rush about it, but affecting the whole world rather than a small particular area. By draining rapidly a limited resource, accumulated geologically over tens or even over hundreds of millions of years, Man has contravened the first of our rules, with the clear threat of making a ghost town of the whole world. The threat will become a reality, unless the technical impetus supplied by the oil can be used to provide us with access to some new and much longer-lasting source of energy.

Use by the United States (ca. 1970) of economically important

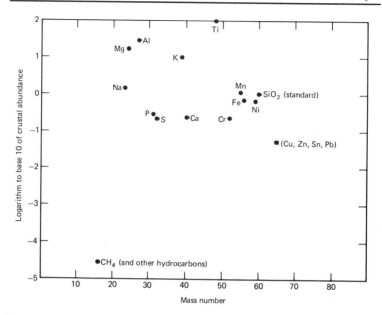

Figure 10   Crustal abundance, in arbitrary units, of several materials as a function of their atomic mass number.

materials is related in Figure 10 to their availability in the Earth's outermost rocks, in the oceans as well as in the air. The aim of this figure is to distinguish plentifully available materials from those that are more rare. The left-hand scale is logarithmic and abundances have been standardized relative to the common material silica (SiO_2). High values indicate ample abundance, low values indicate deficiencies. The critical deficiency is seen to be that for the hydrocarbons—oil, coal, and natural gas.

Another way of displaying this information is shown in Figure 11. This Figure gives the length of time in years for which resources of available materials will last, the left-hand scale again being logarithmic. The estimate of more than 1000 years for the world's coal resources is probably a rather serious over-

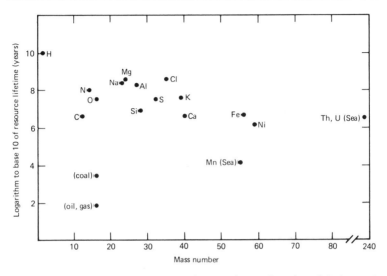

Figure 11   Resource lifetime of several materials as a function of their atomic mass number.

estimate. It assumes all the coal in the ground to be recoverable at zero energy cost. Most of the coal would actually use as much energy or more to recover than it would deliver, at any rate if recovered by present day techniques. In contrast, all other estimates but the hydrocarbons are underestimates—they refer only to the most conveniently recoverable resources. For example, the energy estimate for uranium and thorium refers mainly to recovery from seawater, and does not include the larger quantities of uranium and thorium in the continental granites.

Recovery of uranium and thorium from sea water is highly profitable from an energy point of view. It has been estimated that the extraction of uranium from seawater would cost about $400 per pound, whereas the total energy to be extracted from a pound of uranium has an economic value of some $20,000. The "profit" is thus enormous, besides which seawater is readily available to nearly all the nations of the Earth. Essentially for almost all the materials that we need, the lifetimes are running in the region of millions of years, which are minimal estimates.

Although it is not my purpose in this lecture to enter in any major way into controversial social issues, I think I must say to any of you who hold environmental objections to energy programs based on uranium and thorium that your views are only possible because you happen to enjoy the great luxury of living in the hydrocarbon era. It is generally agreed that this era must be quite short-lived. If no adequate replacement for the hydrocarbons is forthcoming in the future, most of the people then living in the world will die, 4 billion people at today's population level, perhaps 10 billion in the future. Energy from uranium and thorium is the only adequate replacement presently known to be possible.

But to come back to Figure 11, the lifetimes shown here have been calculated for the consumption of materials at 1970 rates. For only the past six years, consumption has moved up appreciably. If the world population were to continue its present out-of-control increase, the lifetimes shown in Figure 11 would soon become meaningless. No conceivable availability of materials can permit a further long-sustained rise in the number of people living in the world. If the present rate of increase were to continue for a further 20 to 30 generations, the human population would become so large that the whole of the Earth's atmosphere could supply a ration of only a single breath to each person. Plainly then, population limitation must come at no

very distant date. It may come forcibly and painfully through overcrowding and material shortages, or it may come through commonsense and voluntary action; by reducing birth rates everywhere to equality with death rates, it could come already before the end of the next year. If this were to happen, nothing concerning the future of our species would then look unfavorable.

The Chaos of
Scientific Cosmology

by

STANLEY L. JAKI
Distinguished University Professor
Seton Hall University

The months which it will take for this lecture to appear in print will bring us to the year 1977. For a historian of cosmology, 1977 marks the 200th anniversary of the death of J. H. Lambert, author of *Cosmologische Briefe über die Einrichtung des Weltbaues* or *Cosmological Letters on the Arrangement of the World-Edifice*.[1] The *Cosmological Letters* is a landmark in the history of cosmology for more than one reason. One of those reasons tells something of the *double entendre* in this lecture's title where the word chaos can be taken either in an objective or subjective sense. In other words, the chaos in question refers not only to that primeval condition of things which is usually taken as the starting point of cosmic evolution. The same word also stands for that chaotic condition which characterizes time and again what is offered in brief comments and extensive studies as well on the history of cosmology.

A good illustration of this can be found in staple references to Lambert's *Cosmological Letters,* although the matter is not altogether inexcusable. In its original German version the full text of Lambert's work can be found even today only in the extremely rare first and only edition of 1761.[2] For many years English-speaking people interested in Lambert's cosmology but unfamiliar with German had to rely on a book entitled *The System of the World,* or the English translation made in 1800[3] of a French digest produced in 1770[4] from the German original. Actually, that English translation was not much help because in addition to being a digest it was as rare if not rarer than the book written by Lambert himself. As one can easily guess, the sequence of digesting cumbersome German into clear French which in turn is transformed into pragmatically hazy English can easily breed that conceptual chaos which is the uncanny equivalent of what the French aptly call the *fausse idée clair,* or the clear but false idea.

Such a clear but false idea made its furtive appearance when relatively brief excerpts from that English translation of the French digest of the German original appeared in the well-known anthology which M. K. Munitz published in 1957 under the title, *Theories of the Universe: From Babylonian Myth to Modern Science.*[5] Munitz failed to inform his readers about what may be called the English digest's "French connection." Most likely he did not provide that information because it would have immediately revealed the questionable value of the excerpts he had offered as Lambert's cosmology. Even worse, he seems to have failed to compare the contents of the excerpts with what is contained in the German original. Otherwise he would hardly have stated in introducing the excerpts that Lambert advocated a single infinite hierarchical universe.[6] What is, then, the reason that Munitz felt that he could dispense with reading the original and attribute to Lambert the idea that the universe was infinite? The answer lies in a *fausse idée claire*, the idea of an infinite Euclidean universe. Among the various clear but false aspects of that universe is the belief that once the seventeenth-century scientific revolution had been completed, for any thinker of any consequence the universe could only appear as infinite in space if not in time.

The actual situation was far from being so simple.[7] Startling evidence can be gathered in that respect from none other than Voltaire whose popular account of Newtonian physics was very influential in making Newton triumph in Cartesian France. In that book, *Elemens de la philosophie de Newton*, which was voraciously read in France in many editions during the middle part of the eighteenth century, Voltaire emphatically claimed that according to Newton the universe was finite.[8] It was only in the nineteenth century that the idea of an infinite universe had become a shibboleth parroted by any and all, although even then there were some telltale hesitations.[9] All this will, of course, be revealed only through a careful reading of contemporary sources, a pastime for which there is no substitute as long as one aims at even a modest measure of scholarship. It is the chronic neglect, and at times the haughty disregard, of those sources which breeds false but clear ideas and will provide one with confidence, for example, to claim that Lambert held the universe to be infinite.

One of Lambert's reasons for holding that the universe was finite was the contradictoriness of an actually realized infinite number.[10] He stated this reason in the Preface of his work and later he once more emphatically reasserted his belief in the necessary finiteness of a universe[11] which, if it was purposeful, had, according to Lambert, to be organized in a hierarchical fashion. Being a good scientist, which Immanuel Kant had never been,[12] Lambert was not to repeat the blunder of Kant who a few years earlier had also proposed a hierarchical but infinite universe in his book, *Allgemeine Naturgeschichte und*

Theorie des Himmels or *Universal Natural History and Theory of the Heavens,* most of which is available also in English under the title, *Kant's Cosmogony.*[13] Only an amateur scientist could argue as he did that while an absolute central point was meaningless in Euclidean space, such a central point implied no contradiction in actual, that is, physical Euclidean infinity. About that infinity and its absolute center, where according to Kant the evolution of stars, galaxies, supergalaxies, and the infinite sequence of ever higher-ranking supergalaxies had started, the best one can say is the definition of infinity given by the proverbial schoolboy: "Infinity is where things happen which don't."

To hint that Kant's cosmogony is a parody of science, or rather a science turned into a conceptual chaos, may seem highly suspect for two reasons. One relates to that single page in Kant's work which has scientific merit where he gives the correct explanation of the visual appearance of the Milky Way.[14] This was a splendid achievement in itself but one in which Kant was greatly helped by what he had read in an essay of Maupertuis about the various shapes of the nebulae and by what he had found in Thomas Wright's speculations about the Milky Way.[15] Moreover, the explanation given by Kant had already been formulated in 1749 by Lambert himself, although he published it only 12 years later in the *Cosmological Letters.*[16] In other words, the explanation was in the air and could be found by such amateurs as Wright and Kant. What should seem rather astonishing is that the explanation in question came so late in the history of science. Well, Newton said not more of the Milky Way than that its whiteness resembled the tail of comets.[17] A curious infinite universe which had no room for the Milky Way, a point invariably overlooked in the cosmological literature dealing with Newton and erstwhile Newtonians.

The other reason which may caution against taking Kant, the cosmologist, lightly has to do with the fact that Kant's cosmology had, as time went on, turned into one of those clear ideas which are part of the intellectual heritage of our times. It has received a special boost by Karl R. Popper, who for many today is the chief interpreter of the history and philosophy of science. Countless is the number of those who have read in his *Conjectures and Refutations* the old truth that ultimately all science is cosmology.[18] But just as many believed his claim made there that Kant had anticipated the cosmology of James Jeans and that therefore he is one of the greatest cosmologists of all times.[19] Not being a good conjecture, Popper's claim does not deserve refutation, but being a widely accepted conjecture it needs a comment or two. The chaos prevailing about Kant the cosmologist has its origin in that halo which Helmholtz placed around him in 1854. Like all halos, it is visible and beautiful only when looked upon from a distance. On a closer inspection both Helmholtz's handling of Kant's cosmogony and Kant's cosmogony itself show the debilitating effects of extrascientific motivations. Helmholtz, as I have recently concluded, was misled by his patriotism,[20] whereas Kant became the victim of his overambitious amateurism in science, a conclusion that had already been reached in 1924 by C. V. L. Charlier, one of the leading cosmologists of the early part of our century. "The *Naturgeschichte des Himmels*," reads Charlier's statement, "is, scientifically, of very small value; that the comparison of it with the planetary cosmogony of Laplace is highly unjust and misleading; also that it cannot be used as a working hypothesis, ... As a *popular* treatise on cosmogony I consider the *Naturgeschichte* of Kant unsuitable and even dangerous as inviting feeble minds and minds uninstructed in natural philosophy [physics] to vain and fruitless speculations."[21]

The comparison with Laplace, which Charlier characterized as unjust, does not mean that Laplace's cosmogony is much better than that of Kant, a fact which makes a rather sad reflection on Laplace. Between 1796 and 1824 Laplace produced five forms of his cosmogony,[22] all gravely and obviously defective, but no scientist of those times cared to comment on this, a curious fact in view of the continual references during that time to Laplace's work in celestial dynamics. Well, some of the scientists in question would have cared to make a comment or two, had they mustered the courage to do so. This was noticed by Olbers, better remembered as an astronomer than as a physician, who spent almost a year in Paris shortly before Napoleon's fatal excursion into Russia. With a clear reference to Laplace's cosmogony Olbers wrote from Paris to Bessel in Königsberg that French men of science (and he listed such illustrious names as Lagrange, Delambre, Poisson, and Arago) bowed before the authority of Laplace by remaining silent on his cosmogony.[23] But they at least refused to endorse it. As late as 1859 Babinet of optics fame could say that Laplace's theory had failed to create any interest in French scientific circles.[24] French interest in Laplace's theory arose only after Helmholtz in Germany and Spencer in England[25] had created the myth of the Kant-Laplace theory and praised it as the greatest feat of human intellect. They and others have invariably failed to study the successive forms of Laplace's theory, which show that Laplace systematically ignored some new discoveries to save his theory[26] and carefully avoided testing its major steps in a quantitative manner,[27] a test which was well within the reach of the mathematical tools available to him and partly produced by him. The ensuing chaos of admiration for Laplace's theory had become so great that hardly anybody paid attention when in 1884 Fouché pointed out before the Academie des Sciences in Paris that on the basis of Laplace's theory one could not account for the distribution of angular momentum in the solar system and therefore the theory must be abandoned.[28] Indeed, sixteen years later, at the turn of the century, that basic defect of Laplace's theory had to be rediscovered by two American scientists, Chamberlin and Moulton.[29]

The nineteenth century should not, however, be left behind without saying something about another chaotic feature of its cosmological speculations. That feature had its origin in the overreaction that had taken place against the closed Aristotelian universe during the seventeenth century when intelligibility became equated with an infinity conceived in a Euclidean fashion. When Lord Kelvin declared in 1884 that infinity was comprehensibility itself but finitude was incomprehensible,[30] he merely voiced an already hallowed clear but false conviction. The conviction had its chaotic ingredients, some of which had already been laid bare in Kepler's arguing against Bruno, in Bentley's correspondence with Newton, and in a quickly forgotten paper which Halley read in 1720 before the Royal Society and which was resuscitated a hundred years later by Olbers. I have devoted a full monograph to the various aspects of that strange story which I called the "paradox of Olbers' paradox."[31] Of those various aspects let me here speak briefly of one general and one particular point both of which reveal something of that chaos in which scientific cosmology can lose itself time and again.

The general point is that it is a mixed blessing for science to take a particular concept, such as Euclidean infinity, and declare it to be equivalent to intelligibility itself. This is equivalent to turning scientific notions into objects of worship, which they are never meant to be. It was in fact for fear of provoking the wrath of the worshipers of Euclidean infinity—who had just received fresh support from Kant—that Gauss declined to devote in full his genius to non-Euclidean geometries and to their application to cosmology. Since I am not one of those historians of science who praise geniuses briefly and attribute at length their accomplishments to sociological and genetic conditioning, my remark should not be taken as a suggestion that Gauss, or Riemann, or Schwarzschild should have anticipated by 100, or by 50, or by 15 years Einstein's general relativity and the cosmologies based on it. What I suggest is that even science, to say nothing of the broader cultural outlook, might benefit by a

modest measure of caution about the presumed absolute validity of some propositions particularly dear to the scientific and
philosophical spirit of the age.

The alternative to that caution is a kind of chaos of which
dichotomy in thinking, bordering at times on schizophrenia, is
an example. In late nineteenth-century cosmology that dichotomy was exemplified in the dispute about island uni-
verses. Its origin was the question whether the thousands of
nebulae sighted by the two Herschels and Lord Rosse were
units comparable in size and mass with our Milky Way or
whether they were much smaller units subordinate to it. The
dispute ended in the general acceptance, during the closing
decades of the nineteenth century, of the view that all visible
galaxies were far smaller than the Milky Way and that they were
distributed on its two sides within a space composed of two
hemispheres of which the Milky Way was the main plane.[32]
This view implied that the infinite realm of matter of stars and
galaxies beyond that imaginary sphere was irrelevant for science because the radiation reaching us from all visible galaxies
was negligible when compared with that of the sun. Such was
the solution which Lord Kelvin gave in 1901 to Olbers'
paradox,[33] a solution which had more to do with his somewhat
chaotic fondness for infinity than with a detached view of the
implications of Olbers' paradox for that very same infinity, insofar as it is physically realized.

Of course, the paradox can be solved if account is taken of
the finite time span during which stars are luminous[34] and of its
ratio to the "age" of the universe. At any rate, a universe constituted of an infinite number of extinguished stars still has its
gravitational paradox, or the consequence that in such a universe that gravitational potential is infinite at any point. The
proof is elementary. It was first given in mathematical form by
Lord Kelvin,[35] then presented 18 years later in a different form
by Einstein in his popular book on relativity.[36] Much less is

known about that consternation with which Einstein's mostly scientific audience received, during his visit to Paris in 1921, his proposal that the notion of infinite mass is something to be avoided in scientific cosmology. The measure of that consternation can be gauged from the fact that as late as 1938 Walther Nernst, a Nobel laureate, still passionately defended infinity as the sole reliable foundation of science.[37] Eddington hardly voiced the scientific consensus when he wrote about the same time: "That queer quantity 'infinity' is the very mischief and no rational physicist should have anything to do with it."[38]

To realize the measure of mischief which infinity is capable of doing in physics, it should be enough to think of the "infinity catastrophe" that had wrecked explanations of blackbody radiation before Planck eliminated the infinitely small by introducing the quantum of action. Also, a brief recall of the need of boundary conditions when working with differential equations should be enough of a reminder that one should tread very carefully when infinity arises on one's scientific horizon. This is not to suggest that infinity in the physical world is an unqualified chaos, but it seems that it can be dissipated only at a price which has a metaphysical tag on it. The tag in its obvious appearance is merely physical, though with a strong psychological connotation. That psychological component derives from the fact that the singularity implied in the tag is most unnatural in the sense of being most unexpected. What appears most natural to our sense perception is plain three dimensionality, and this is why it is so natural to see in Euclidean infinity the necessary form of the entirety of nature, that is, the universe. The curvature of such a universe is zero. It is an uncanny symbol of the invalidity of such a universe, unfortunately, that zero, which of all symbols appears to be the most simple, is also the most general and most natural. This is why the notion of an infinite Euclidean universe could parade for three centuries as the necessary form of cosmic existence and prompt so many men of science and philosophy to conclude, without further ado, that the universe is what it appears to be and cannot be anything else.

It should now be easy to understand the shock experienced by many when it was realized under the impact of Einstein's general relativity that a universe with zero curvature is to be ruled out. Clearly, if the universe has an overall curvature which is not zero but say 0.85 or 1.79, then it becomes difficult to rush to the conclusion that the universe is necessarily what it is and cannot be anything else. The specific curvature of the universe acts therefore as what I would call a strong singularity. It forcefully brings into focus questions about the universe, questions which will hardly be perceived by those who with Paul Carus and other monists see in zero the factor which makes everything equal[39] and also turns the universe into that happy place where there is no need for further questions. Once, however, that zero is spoken of with Halsted as the factor which is "coining the Nirvana into dynamos,"[40] even a universe with zero curvature will cease to be that happy universe in which no further questions arise about its ultimate origin.

The relatively recent ability to assign an overall curvature to the universe is an achievement which gave for the first time a truly scientific character to cosmology. The achievement is all the more scientific because it gives on a cosmic scale a hold on that very change which is the source of our perceptions and which it is the very objective of science to describe in a quantitative manner. It is more than a linguistic happenstance that infinitesimal calculus, that great breakthrough in science, had first been called the science of fluxions, or the science of nature insofar as it is in a flux, that is, in a state of change. Of course, change insofar as it is a prompting for meaningful reflections implies the permanence of something during the very process of change. Such a permanent feature of the universe is its curvature. The observed expansion of the universe is written in its curvature, a stark singularity which is given from the very inception of that expansion. It is here that we come face to face with the most telling characteristic of that physical chaos which alone is useful for scientific cosmology. That chaos is anything but a chaos in the customary sense of the word. It is anything but an entity in which everything is mixed up in a wholly chaotic condition. The primeval physical chaos, which is dealt with in scientific cosmology, is really a cosmos and this is the heart of the second meaning of the word chaos intended in the title of this lecture. The Greeks of old spoke of the universe as beautiful, or cosmic, because it appeared to them to have a singular beauty, a beauty singularly structured and proportioned. Such a singular beauty is the so-called primeval chaos out of which the present condition of our universe has evolved. Unfortunately, in order to see convincingly the extraordinary beauty or rather orderliness of that primeval chaos, better called primeval cosmos, one must be able to read the forbidding technicalities of relativistic cosmology, a privilege reserved only for a few.

Yet experts on occasion give a glimpse of that primeval chaos which should make us hold our breath, as if we were placed on the edge of that chasm of which Hamlet once spoke as the gap between "to be" and "not to be." Such a glimpse was provided a little over a year ago by Professor Lovell, whose presidential address given before the British Association for the Advancement of Science made even local newspapers take note and report word for word some of his utterances. The breathtaking point relates to the cosmic background radiation which had been measured in 1973 and 1974 with highly increased precision. That measurement gives an estimate of the fall of temperature during the very first minute of the expansion of the primeval fireball. The change in temperature accounts in turn for a very specific rate of interaction between protons and neutrons during the first minute of cosmic history. "It is an astonishing reflection," stated Professor Lovell, "that if the interaction were only a few percent stronger, then all the hydrogen in the primeval condensate would have turned into helium in the early stages of expansion. No galaxies, no stars, no life would have emerged. The result would be a universe forever unknowable by living creatures."[41]

The drama of being placed on the line separating the actual structure of cosmos, which implies life, from all other forms that preclude life is certainly worth pondering. It is equally worth recalling that in the cosmic drama, as in any great drama, the actual outcome is subtly contained in the very first act. This means, to speak the language of scientific cosmology, that the primeval chaos is actually the primeval cosmos, an entity magnificently ordered and beautifully proportioned in all its parts and in its entirety. The orderliness is simplicity itself, though not in a trivial sense, for the orderliness in question is that of the realm of the so-called fundamental particles, a realm that should appear anything but trivial in any of its characteristics and as such permits no trivial portrayal. On my part I would like to add only such remarks as are of the domain of a philosopher and historian of science. The first remark is that, as Oppenheimer once facetiously noted, our most reliable guess about fundamental particles is that none of them is fundamental. In other words they point to deeper foundations. Whether these deeper foundations will be further layers of particles is largely irrelevant. After all, even our presently known fundamental particles are not particles in the ordinary sense of that word. But whether particles or not, they are singularities and as such they can only be traced back to other singularities. My other remark concerns the derivation of actual physical singularities from a priori general considerations, an effort which should seem suspect from the logical viewpoint and which has so far met with no scientific success. The most monumental of recent efforts of that kind can be studied in Eddington's *Fundamental Theory*,[42] a theory pivoted on the fine-structure constant almost equivalent of 1/137. Eddington considered it so universally fundamental that he looked in cloakrooms for the peg

numbered 137 on which he preferred to hang his hat.[43] Whatever his success in this respect, he signally failed to hang the
universe on that number, which he tried to derive from a priori
considerations relating to epistemology. One excuse for Eddington is that he worked out his theory in the pre-Gödel era, a
statement which needs some explanation because Gödel's incompleteness theorem was proposed in 1931,[44] whereas Eddington's *Fundamental Theory* was published in the mid-1940s.
Gödel's theorem did not sink into academic consciousness until
the mid-1950s and it still has to sink considerably deeper. As
Popper recalled, during the 1930s and 1940s Carnap tried, but in
vain, to make his fellow logical positivists, who dominated the
scene more and more, aware of the fundamental importance of
Gödel's theorem.[45] Today, there is no excuse for ignoring it. The
theorem states that a nontrivial arithmetic system cannot have
its proof of consistency within itself. A brief recall of the complex, at times extremely complex, mathematical underpinnings
of cosmological theory dealing with the primeval chaos of fundamental particles should make it now clear why an a priori
approach to cosmology is precarious. The attempt to declare on
a priori mathematical grounds what the world ought to be cannot have its proof of consistency within itself as far as the
mathematics is concerned. Recourse to experimental evidence
will be of no real help because there is no way of telling, at a
given stage of physical inquiry, whether or not everything that
can be known about the universe has already been learned,
especially in reference to so-called fundamental particles.

Fundamental particles constitute the primeval chaos no less
than they constitute our present-day cosmos. This is what gives
to that chaos a genuinely scientific character, and this is what
makes ultimately possible a scientific cosmology. Our knowledge of fundamental particles and fundamental constants is indeed so far from being chaotic that, as Professor Weisskopf
showed in an article published in *Science* two years ago,[46] once
those constants are given, it is possible to predict that the size

of stars must fall between very specific limits, that our atmosphere must be blue, and that even our highest mountains cannot be much higher than they actually are. Such an achievement reflects by its precision something of that cosmic beauty which should seem the deepest characteristic of the primeval chaos, which is really the primeval cosmos.

In this lecture I have started with Lambert's *Cosmological Letters*, a work which I have described as a landmark in the history of cosmology. I still have to give another reason why it is a landmark. It is such because it is the last cosmological discussion in which the nonevolutionary character of the universe is advocated. Lambert disliked evolution because it seemed to him a harbinger of democracy.[47] He felt very much at ease in the absolutist and hierarchical Berlin of Frederick the Great. Lambert also disliked evolution because he wanted every planet and comet to be the safe and permanent abode of life. It was his preoccupation with life everywhere in the universe which turned his cosmology into a cosmic fiasco, a fate which—if I may remark—has played havoc and still can play havoc with evolutionary cosmologies when written with the a priori aim to find life in every nook and cranny of the universe.[48] Even worse, many of these recent expositions of evolutionary cosmologies are starting from the assumption that there is no purpose. That a nonpurposeful and valueless beginning should ultimately give rise to beings with purpose and values should seem indeeed remarkable. At any rate, as Whitehead had already remarked in 1929, people who devote their lives to the purpose of proving that there is no purpose, constitute an interesting subject of study.[49]

Obviously, a primeval chaos which is a cosmos, a beautifully and singularly structured unit that obviously points beyond itself, is hardly an entity which radically preempts the possibility of purpose. But my lecture is not about the chaos of metaphysical cosmology which deals with purpose, but of scientific cosmology which is about quantities and not about values and about purpose. As I tried to show, the chaos useful for scientific cosmology is a structure permeated everywhere with the most specific quantitative singularities. Herein lies the reason why the various kinds of chaos that had been proposed prior to our times as starting points of cosmogonical processes proved to be inadequate. They all lacked quantitative singularities, singularities underivable from general considerations. The best known of these not sufficiently scientific chaoses, Laplace's nebula, was proposed a generation after Lambert's *Cosmological Letters,* but some others had been proposed before, and a brief remark about them will not be out of place. As to the chaoses proposed in classical antiquity, they were of three main kinds. One was Heraclitus' chaos, which is well known through a long passage from it quoted in Popper's *Open Society.*[50] That chaos of Heraclitus is based on the radical avoidance of any and all consistency. Clearly such a chaos is of no help for science which is inconceivable without consistency or laws. The second kind of chaos is that of Leucippus which had a rudimentary scientific merit precisely because he not only assumed atoms but also endowed them with little hooks, so many quantitative singularities, which could not be derived from the notion of atoms but which were indispensable to make the atoms useful for cosmic evolution. The third kind of chaos, proposed by Democritus, reveals something of his greatness as a philosopher insofar as such greatness is measured by the philosopher's consistency. Democritus was resolved not to accept any kind of universally valid singularity or limitation in the universe. This is why he claimed that atoms could not be limited to a given size. On such a basis, one can build philosophies and even worlds but hardly a scientifically meaningful and investigable universe.

When in modern times Descartes first proposed an evolutionary cosmology starting with a chaos, he referred to the chaos of the ancients, but not in order to borrow anything from them. He was also determined to make the primeval chaos as primitively simple as possible. Ultimately that chaos was sheer three-dimensional continuous extension onto which he tacked only one singularity: the large domains with a diameter equivalent to the average distance of stars. It was from that singularity that he tried to derive everything else in the actual world, an effort doomed to failure precisely because he wanted to get more out of his original presupposition than what it contained. Something which in particular he failed to obtain was light. The rays of light which he pictured as so many rigid continuous rods could but offset their effect at any point. The universe of Descartes was enveloped in darkness in an ironical rebuttal of the title, "The World, or a Treatise on Light," which Descartes gave to the first form of his account of the universe.[51]

But there is an even more striking aspect to Descartes' cosmology, an aspect which derives precisely from its being an evolutionary one. Moreover, that aspect gives to it a strikingly modern character. The character in question is the ultimate darkness or death which is in store for the universe, in agreement with the three laws of thermodynamics formulated about a hundred years ago by Helmholtz, Kelvin, and Clausius. In a popular but still valid form those laws can be stated as follows: you cannot win, you cannot break even, and you cannot even get out of the game. In the 1860s and 1870s the ultimate heat death of the universe was avidly discussed, but by the end of the century interest in the topic diminished considerably, partly under the influence of Mach's empiricist philosophy according to which scientific questions relating to the whole cosmos are meaningless, because the cosmos as such is not observable. Mach, if it may be noted in passing, was very suspicious of cosmology. He called it a Greek mania[52] and would have strongly disavowed having his name attached to what is spoken of in modern cosmology. Although Mach rightly saw the source of science in man's urge to achieve an ever more comprehensive view of the world,[52] he slighted cosmology and would have strongly disavowed general relativity precisely because it implied the validity of the notion of the totality of physical things, or the universe. At any rate, if the so-called Mach's principle had a clear origin in Mach's writings, it would not have the variety of meanings suggestive of conceptual chaos that it has today.[53]

By the time of Mach's death in 1916 the topic of the heat death of the universe had become so unfashionable that Eddington and Jeans produced quite a shock by bringing it back into the center of attention. In fact Eddington felt it necessary to defend himself with the remark that some, oddly enough, had found the scientific equivalent of the expression, "heaven and earth shall pass away" quite unorthodox.[54] Indeed, quite a few people, not only scientists but also some theologians of those theologically very liberal times, seemed to believe that if not they, at least the universe was going to last forever. This is not to suggest that no new stars can form out of the ashes of supernovae. What is suggested is that in the stellar birth-growth-decay-rebirth cycle energy is expended in a way which prevents the universe from becoming a perpetual motion machine. There are some who prefer to see in the universe such a machine, a curious preference based on an even more curious if not chaotic logic. For the very meaning of logic is at stake when a property such as perpetual motion, which consistently and for very good scientific reasons is denied to any part, natural or artificial, of the physical universe, is blandly and without any good scientific reason attributed to the totality of things physical, that is, the universe. Logic rather suggests that just as mankind will not last forever, the activity of the starry realm too will come to an end. Not only along the parameter of space but also along the parameter of time, the universe seems to bear the mark of a singularly strong limitation.

In that latter connection too Descartes was a prophet without suspecting it. He started with a universe which originally had contained but stars, at the center of each vortex or domain. Because these vortices were not absolutely equal, and because he wanted to account for the novae by assuming that the stars could be covered by a hard crust, lose that crust again for a while, and then finally become encrusted for good, it was inevitable that the original balance between neighboring vortices be upset. This resulted in the capturing of an encrusted star by a still live star. In other words, in the Cartesian evolutionary cosmology, all the planets and all the comets were former stars. The planets were permanently captured dead stars, while the comets were occasional interlopers into any solar system. The ultimate fate of the Cartesian universe was a single star surrounded by an infinite number of dead stars or planets and comets, and finally even that star was to be encrusted. All this was implied in Descartes' cosmogony, but he never spelled it out, nor was it pointed out either by his admirers or by his critics,[55] Newton being one of the chief among the latter.

The darkness which enveloped the Cartesian universe is the result of Descartes' presumptuous thinking about the original chaotic form of the universe. But even a scientific cosmology which acknowledges the extreme complicatedness of the original chaos can lead to the conclusion that light in the universe cannot forever be taken for granted. The black hole into which the universe may eventually turn is a gigantic warning in that respect, but there are many smaller pointers worth pondering. Very recent observations of neutrino emissions from the sun suggest that the stability of the sun as a light source is much more precarious than it has been believed. Clearly, the observed universe, which in a broad sense is a universe of light, is a very peculiar, singular entity. It is like a flash across an infinite span of the unobservable or darkness, a flash which does not explain itself. To be sure, the words of Genesis, "Let there be light," were not meant to be an instruction in scientific cosmology. But like all poetico-prophetic words, they have an instructiveness far beyond their directly intended meaning.

By that instructiveness I especially mean the deepest which the study of cosmology can offer. About that study let me first note that the word cosmology is a relatively new word. It first appeared in the title of a book only at the beginning of the eighteenth century and it was only a little over 200 years ago that an encyclopedia carried an article on cosmology. The encyclopedia was none other than the one conceived and edited by Diderot and d'Alembert. It fell to the latter to write the article on "cosmologie" which culminates in the following statement: "The main profit we should derive from cosmology is to raise our minds with the help of the general laws of nature to its author whose wisdom has established those laws, allowed us to perceive those that are needed for our benefit or for our entertainment and has hidden from us the rest to make us learn the art of doubting."[56]

The validity of these words is amply attested by the history of cosmology. Its often chaotic advances should help keep a healthy measure of doubt about its latest gossips, however fashionable and sophisticated. Its spectacular feats certainly match the best entertainment. As to our chief benefit which derives from cosmology, let it merely be recalled that all objective truth rests ultimately on the truthfulness of the cosmos, a truthfulness which is objective only if we find it given by the cosmos rather than given to it by our proclivity to a priori thinking. This proclivity, as shown by the history of post-Cartesian and especially of post-Kantian philosophy, has become the chief source of subjectivism instead of securing objectivity. Subjectivism not only makes cosmological discourse impossible, but in its so-called scientific form, or behaviorism, it does not even make it possible to tell a terrorist from a freedom fighter,[57] or to pass a valid judgment on corrupt police and corrupt politicians.

A great loss to be sure even if not considered with relevance to cosmology, for judging politicians has become a craving of modern man, a point of which Watergate has reminded all of us. Before Watergate policemen in New York were the favorite target of the self-righteous, who were ready to forget about the

ultimate cosmic ground which alone can make the self right.
Forgotten it certainly seemed to be during that dinner conver-
sation of which I was part with a few other philosophers and
scientists and which had for its topic police corruption in New
York and elsewhere. My reticence made me conspicuous to the
host, a towering figure of twentieth-century physics, who put
me on the spot with the question: "But, Stanley, don't you
agree that policemen should behave like angels?" I waited
briefly and said, "Hell, no!" At this all knives and forks
dropped. "What do you mean?" I was asked in reply. "Well,
look," I said, "logic tells me that a society which does not believe in angels has no right to policemen that behave like
angels."

Of course, by angels I did not and do not mean the ones that
allegedly dance on pinheads and offer themselves to a heedless
headcount. By angels I mean truths and norms valid for all
thinking beings, whatever their temporal and social param-
eters. A cosmology which knows the difference between subjective chaos and objective cosmos is one of the greatest
achievements of such beings and also their greatest safeguard
against intellectual, ethical, and behavioral relativism which
ultimately leaves everything to the caprice of the fleeting moment, or rather to the so-called empirical which is always
momentary when severed from an objective and metaphysical
perspective. That such is the case is borne out by the cosmological dicta of the protagonists of the empirical movement,
Hobbes, Hume, Diderot, Comte, Mill, Mach, James, and beyond.[58] They secured to empiricism the dubious distinction
of holding high the possibility of a universe coherent only in
some but not in all its parts and coherent only for one cosmic
age but not for another. Needless to say, this way of thinking
about the universe defeats the rationale of scientific cosmology,
especially as we have it since Einstein.

Tellingly enough, those protagonists of empiricism were fully
aware of the fact that a coherent cosmos, as long as it is empiri-

cally given and not determined a priori, invariably becomes the stronghold of theists. No less tellingly, the agnostic Einstein also came to recognize that the view of the objective cosmic order he unfolded in a scientifically consistent manner could easily make his positivist friends think that he had fallen, as he put it, "into the hands of priests."[59] What he meant was that the cosmology based on general relativity had that objective and nonrelative character which from times of old has always been assigned to the cosmos by those who, precisely because of this, saw it as the handiwork of the Creator.

Their first utterances should seem chaotic by modern standards. Nor was anything radically original in that "darkness that lay upon the face of the deep," or the primeval chaos. The phrase had for its source the general Semitic folklore codified in the *Enumah Elish*.[60] But whereas the *Enumah Elish* offered a picture of the universe in which order could never permanently prevail over chaos—the two alternated in never-ending cycles—the compiler of Genesis codified a totally different cosmic perspective. In it there was no room for the slightest doubt that the order and structure imposed on that chaotic darkness that "lay upon the face of the deep" were to turn into chaos again.[61] The difference between these two perspectives meant the difference between the repeated stillbirths of science in all ancient cultures and its only viable birth in a culture, medieval Europe, which was the first broadly based culture steeped in the cosmological perspective about the chaotic beginning as given in Genesis. Such is the link that ties the history of science, or cosmology (since all science is ultimately cosmology), to the objective truth embodied in the cosmos.

Indeed, the more that is unfolded by scientific cosmology about the "chaotic" beginning, the more it appears a cosmos in embryo, that is, the most objectively rational entity and event conceivable by human intellect. It is objectively rational precisely because the rationality in question is not a subjective a priori product of human intellect, but points beyond itself to a

creative reason transcending all the cosmos. To avoid that inference one can, for instance, take the stance of an agnostic with
regard to ultimate questions, a stance that cannot be held consistently because once the *ultimate* question has been disre-
garded, the next-to-last question becomes the ultimate though
inconsistent question. About that process it should be remembered that just as there is no logical recourse to infinite regress,
there is no such thing as infinite retreat. Another way to avoid
the inference in question is to renounce the thesis that scientific
cosmology is about the totality of consistently interacting entities, the very thesis that lifted cosmology through Einstein's
general relativity to a level which is genuinely scientific. It is
also the level which Kant wanted to discredit as rational and
scientific, because he was fully aware that once the totality of
things, the universe, had been accepted as an objectively valid
notion there was no way of barring the inference to the Maker
of that universe.[62]

Still another way, and a very modern one, to avoid that inference is to introduce into scientific cosmology spectacular
sleights-of-hand that fly in the face of logic and science. Their
scientific merit, whatever their mathematical garb, does not
exceed the merit of the concave ether walls which Rankine
conjured up in cosmic space a hundred years ago and without
the parade of mathematics.[63] With the help of those imaginary
ether walls Rankine hoped to refocus the dissipated energy of
stars and secure thereby the character of a perpetual motion
machine to the universe. About the logic of such and similar
efforts one need not even ponder the logic that nothing can give
rise to but nothing. It is enough to recall that one's lifting of
oneself by one's own bootstraps is still to be demonstrated.[64]
Claims to the contrary are but an invitation to chaos with respect to logic as well as to science. In brief, it is but a short step
from that beautiful primeval chaos which cosmology is supposed to deal with to a shabby chaos in which cosmology, as its
history shows, can easily lose itself. In that loss of thinking

about the cosmos, the thinking man's fate and fortunes are vitally involved and therefore the loss should seem even to a noncosmologist as one with stakes too high to be taken lightly. The history of cosmology is composed of pages, some of which are of more than purely historical interest, and which warn against taking the original chaos for anything but a cosmos, lest cosmology and other branches of learning turn into a subtly chaotic discourse.

NOTES AND REFERENCES

1. Published only a week before the delivery of this lecture in my translation with introduction and notes (New York: Science History Publications; Edinburgh: Scottish Academic Press, 1976).

2. Augsburg: bey Eberhard Kletts Wittib.

3. Translated from the French by James Jacque, Esq. (London: printed for Vernor and Hood by J. Cundee, Ivy-Lane).

4. The digest, *Système du monde* (Bouillon: aux dépens de la Société Typographique), was the work of B. Merian, an admirer of Lambert and also a member of the Berlin Academy. The title page of the second edition, *Système du monde* par M. Lambert; publié par M. Merian (Seconde édition; à Berlin, et se vend à Paris . . . et à Genève . . . , 1784), simply ascribed to Lambert the work about which the Preface of the first edition stated that it was merely a condensation of the original.

5. New York: The Free Press; since 1965 available in paperback edition.

6. "A similar speculative cosmology, making use of the notions of Newtonian mechanics and the idea of a hierarchy of celestial systems culminating in a single infinite universe, was worked out by Johann Lambert, a contemporary of Kant, in his *Cosmological Letters*, along lines that are in some respects strikingly similar to those of Kant" (ibid., p. 145). A universe, needless to say, which is infinite, can only be single.

7. Little if anything is revealed of this in S. Bochner's article, "Infinity" in *Dictionary of the History of Ideas* (New York: Charles Scribner's Sons, 1968–1974), Vol. 2, pp. 604–617, where it is claimed that in the three editions of the *Principia* Newton avoided committing himself on the infinity or finiteness of the universe and used rather the Copernican term "immense," "although between the first and second editions, in a written reply to a query from the . . . intolerant divine Robert [Richard] Bentley,

Newton made some kind of 'admission' that the universe might be infinite" (p. 608). The fact was that in that correspondence Newton seemed to defend the infinity of the universe and in spite of Bentley's perfectly valid argument that in an infinite universe of homogeneously distributed stars the gravitational attraction was at any point infinitely great from any direction and that therefore the net result was zero. This argument of Bentley's was overlooked by A. Koyré, whose *From the Closed World to the Infinite Universe* (1957; New York: Harper and Brothers, 1958, pp. 178–179) was referred to by Bochner as the authoritative treatment of the question.

8. The *Elemens* was first published in 1738. For the passage, "car si selon Newton (et selon la raison) le monde est fini . . . ," see *Oeuvres complètes de Voltaire* (Paris: Garnier Frères, 1877–1885, Vol. 22, p. 403.

9. The two Herschels were conspicuous in that respect.

10. *Cosmological Letters*, p. 47.

11. Ibid., p. 125. The finiteness of Lambert's universe was also implied in the extremely large though finite mass of the highest-ranking dark body, or "Regent," which kept by its gravitational attraction all the subordinate systems of galaxies and stars in order.

12. A point that cannot be repeated often enough in view of the continued references to Kant as being highly competent in science. A recent example of such presentations is the article "Kant" by J. W. Ellington in *Dictionary of Scientific Biography* (New York: Charles Scribner's Sons, 1970), Vol. 7, pp. 224–225 and 230–231, where, as usual, M. Knutzen is given as Kant's private teacher in the sciences, in complete disregard of the fact that Knutzen was not a scientist. The courses he taught at the University of Königsberg were on philosophy, ethics, and religion; of his many publications only one, on the famed comet of 1744, had some scientific connotation. Such and similar details can easily be gathered from the standard monograph on Knutzen by B. Erdman, *Martin Knutzen und seine Zeit: Ein Beitrag zur Geschichte der Wolfischen Schule und insbesondere zur Entwicklungsgeschichte Kants* (Leipzig: Verlag von Leopold Voss, 1876; reprinted, Hildesheim: H. A. Gerstenberg, 1973). For further details on Kant, the "scientist," and on the myth about him, see my *Planets and Planetarians: A History of Theories of the Origin of Planetary Systems* (Edinburgh: Scottish Academic Press, 1978), pp. 146–148, and Lecture Eight: "Arch without Keystone," in my Gifford Lectures, *The Road of Science and the Ways to God* (Chicago: University of Chicago Press, 1978).

13. In the translation of W. Hastie (Glasgow: James Maclehose and Sons, 1900), reprinted in part three times during the last 10 or so years. Being an admirer of Kant, Hastie preferred not to translate the third part in which Kant discoursed with free abandon on the physical and moral characteristics of denizens of other planets, characteristics which he felt entitled to ascertain from the analysis of their relative distances from the Sun! This third part is being published in my translation as an Appendix in *Cosmol-*

ogy, History of Science and Theology, edited by W. Yourgrau and A. D. Breck (New York: Plenum Press, 1977), pp. 387–403.

14. *Kant's Cosmogony,* p. 55. Kant's identification of other nebulae as other Milky Ways, or systems of stars, should be seen in its context (p. 63) where he envisions an infinite system of ever higher ranking galaxies, an idea which could but appear visionary even in its time with little or no appeal for properly scientific discussion. As documented in detail in my *The Milky Way: An Elusive Road for Science* (New York: Science History Publications, 1972), pp. 198–202, the few valuable details in Kant's cosmological discourse did not create an echo until Herschel observed them through his giant telescopes with no influence from Kant.

15. See *The Milky Way,* pp. 196–197 and 199–200; also see *Cosmological Letters,* pp. 23–24.

16. As Lambert himself informed Kant on this point in a letter of Nov. 13, 1765; see *Cosmological Letters,* p. 8.

17. The "uncertain agitation" of the tails of comets may be due, Newton reasoned, to "parts of the Milky Way which might have been confounded with and mistaken for parts of the tails of the comets as they passed by." See A. Motte's translation of the *Principia,* revised and annotated by F. Cajori (Berkeley: University of California Press, 1962), p. 525.

18. New York: Harper and Row, 1968, p. 136.

19. Ibid., pp. 175–177.

20. See my *Planets and Planetarians,* p. 136.

21. "On the Structure of the Universe" (Hitchcock Lectures at the University of California, April 1924) in *Publications of the Astronomical Society of the Pacific,* **37** (April 1925), p. 63.

22. See my "The Five Forms of Laplace's Cosmogony," in *American Journal of Physics,* **44** (1976), pp. 4–11.

23. In a letter of July 10, 1812; see *Briefwechsel zwischen W. Olbers und F. W. Bessel,* edited by A. Erman (Leipzig: Avenarius & Mendelssohn, 1852), Vol. 1, p. 337.

24. "Cosmogonie de Laplace," in J. Babinet, *Études et lectures sur les sciences d'observation et leurs applications pratiques* (Paris: Mallet-Bachelier, 1865), Vol. 7, pp. 105–106.

25. Spencer's discussion of the nebular hypothesis took up three-fourths of a review of his "Recent Astronomy and the Nebular Hypothesis," in *Westminster Review,* **70** (July 1858), pp. 185–225, of works of Laplace, Arago, the younger Herschel, and Loomis.

26. While Laplace was eager to report the discovery by Herschel of four satellites of Uranus, he kept silent on the retrograde motion of two of them, a feature not derivable from his theory.

27. The principal of these steps was the detachment of rings from the rotating

solar nebula at the actual planetary distances and their respective coalescence into one planetary body.

28. "Sur la condensation de la nébuleuse solaire dans l'hypothèse de Laplace," *Comptes rendus*, **99** (1884), pp. 903–906.

29. For details of the origin, development, and shortcomings of their planetesimal theory, see my *Planets and Planetarians*, Chapter VII, "Collisions Revisited."

30. "I say *finitude* is incomprehensible, the infinite in the universe *is* comprehensible . . . . What would you think of a universe in which you could travel one, ten, or a thousand miles, or even to California, and then find it come to an end? Can you suppose an end of matter or an end of space? The idea is incomprehensible." Such was Kelvin's defense of the exclusive intelligibility of Euclidean infinity in his lecture, "The Wave Theory of Light" in *Popular Lectures and Addresses* (London: Macmillan, 1889–1894), Vol. 1, p. 314, in which no reference was made to the possible physical realizations of non-Euclidean geometries for some time under discussion.

31. *The Paradox of Olbers' Paradox: A Case History of Scientific Thought* (New York: Herder & Herder, 1969).

32. The idea was given a graphic illustration in Newcomb's *Popular Astronomy*, a standard textbook during the closing decades of the nineteenth century. For a variation of that illustration, see my *The Milky Way*, p. 343.

33. "On Ether and Gravitational Matter through Infinite Space," in *Philosophical Magazine*, **2** (1901), pp. 161–177.

34. Such a solution is given by E. R. Harrison, "Why the Sky Is Dark at Night," in *Physics Today*, **27** (Feb. 1974), pp. 30–36.

35. Using Green's theorem (1828) for force at a surface of any shape obeying the inverse square law, Kelvin obtained for a spherical surface S the formula $Q = 4\pi/3r\rho$, where Q denotes the mean value of the normal component of the gravitational force at any point of S, and ρ is the mean density of matter enclosed within S of radius r. "This shows," Kelvin concluded, "that the average normal component force over the surface S is infinitely great, if ρ is finite and r is infinitely great." See "On Ether and Gravitational Matter through Infinite Space," pp. 168–169.

36. In speaking in his *Relativity: The Special and General Theory* (1916; 15th edition, 1952; New York: Crown Publishers, 1961) of the cosmological difficulties of Newtonian theory, Einstein noted that according to it "the stellar universe ought to be a finite island in the infinite ocean of space" (p. 106) and offered as proof the following footnote: "According to the theory of Newton, the number of lines of force which come from infinity and terminate in a mass m is proportional to the mass m. If, on the average, the mass density ρ_0 is constant throughout the universe, then a sphere of volume V will enclose the average mass $\rho_0 V$. Thus the number of lines of force passing through the surface F of the sphere into its interior is pro-

portional to $\rho_0 V$. For unit area of the surface of the sphere the number of lines of force which enters the sphere is thus proportional to $\rho_0 V/F$ or to $\rho_0 R$. Hence the intensity of the field at the surface would ultimately become infinite with increasing radius R of the sphere, which is impossible."

37. As reported of W. Nernst's reaction to the measurement of the "age" of the universe on the basis of radioactive decay in C. F. von Weizsäcker, *The Relevance of Science* (New York: Harper and Row, 1964), p. 151.

38. *New Pathways in Science* (Cambridge: University Press, 1934), p. 217.

39. "Infinity is the land of mathematical hocus pocus. There Zero is the magician king . . . . Here all ranks are abolished, for Zero reduces everything to the same level one way or another. Happy is the kingdom where Zero rules!" ["Logical and Mathematical Thought," in *The Monist*, **20** (1909–1910, p. 69].

40. Quoted in B. L. Van der Waerden, *Science Awakening*, translated by A. Dresden, with additions of the author (New York: John Wiley & Sons, 1963), p. 56.

41. The passage quoted is part of the report in the *Evening Bulletin* (Philadelphia, Aug. 28, 1975, p. 60, cols. 6–7) and has been retained without change in the expanded form of the address, "Whence?," in the *New York Times Magazine*, Nov. 9, 1975; for quotation, see p. 88.

42. The title chosen by Sir Edmund Whittaker, editor of the partially completed sixth manuscript, published in late 1946, two years after Eddington's death. The main ideas of an otherwise forbiddingly technical book are well presented in Chapter 11, "Fundamental Theory," in *The Life of Arthur Stanley Eddington* by A. Vibert Douglas (London: Thomas Nelson and Sons, 1957), pp. 145–182. For a more concise summary, see J. Singh, *Great Ideas and Theories of Modern Cosmology* 1961; New York: Dover, n.d.), pp. 168–180.

43. *The Life of Arthur Stanley Eddington*, p. 146.

44. K. Gödel, *On Formally Undecidable Propositions of Principia Mathematica and Related Systems*, translated by B. Meltzer, with an introduction by R. B. Braithwaite (Edinburgh: Oliver and Boyd, 1962). *Gödel's Proof* by E. Nagel and J. R. Newman (New York: New York University Press, 1960) is a paraphrase of Gödel's train of thought for the educated layman. For the cosmological implications of Gödel's theorem, see my *The Relevance of Physics* (Chicago: University of Chicago Press, 1966), pp. 127–130.

45. K. R. Popper, *Conjectures and Refutations* (New York: Harper and Row, 1968), p. 270.

46. "Of Atoms, Mountains, and Stars: A Study in Qualitative Physics," in *Science*, **187** (1975), pp. 605–612.

47. *Cosmological Letters*, p. 170.

48. As illustrated by the long history of devising explanations for the evolu-

tion of our planetary system, and documented in my *Planets and Planetarians*.

49. *The Function of Reason* (Princeton: Princeton University Press, 1929), p. 12.

50. *The Open Society and Its Enemies* (Princeton: Princeton University Press, 1950), p. 15: "The cosmos, at best, is like a rubbish heap scattered at random." As other quotations there from Heraclitus show, he too realized that a complete chaos was unworkable, if not unthinkable, a point which is often overlooked even in twentieth-century discussions about chaos and chance and which was noted with memorable force by the biologist W. R. Thompson almost half a century ago: "A Universe of pure chance is, in the strict sense of the word, unthinkable, by which we mean, not simply something distasteful or dissatisfying, but something on which the mind cannot take hold at all. A world of pure chance is simply chaos, or absolute disorder, and the concept of absolute disorder has no positive intelligible content." *Science and Commonsense* (London: Longmans Green Co., 1937), p. 218.

51. For further discussion of this aspect of Descartes' cosmology, see my *The Paradox of Olbers' Paradox*, pp. 43–44.

52. But see also his *The Science of Mechanics*, translated by T. J. McCormack from the ninth German edition (6th ed.; La Salle, Ill.: Open Court, 1960), p. 556.

53. As documented in H. Goenner, "Mach's Principle and Einstein's Theory of Gravitation," in *Ernst Mach: Physicist and Philosopher. Boston Studies in the Philosophy of Science. Volume VI*, edited by R. S. Cohen and R. J. Seeger (Dordrecht: D. Reidel, 1968), pp. 200–215. "There is no Mach Principle in Mach's writings," wrote J. Bradley in his *Mach's Philosophy of Science* (London: The Athlone Press, 1971), p. 145.

54. *New Pathways in Science* (Cambridge: University Press, 1934), p. 59.

55. See my *Planets and Planetarians*, pp. 46, 57, and 63.

56. "Cosmologie," in *Encyclopédie ou Dictionnaire raisonné des sciences*, Vol. IV (Paris: chez Briasson, 1754), p. 294.

57. At the time, shortly before noon, when I reached this part of my address, many in the large audience were already thinking of the second debate between Ford and Carter to be televised late in the evening. I am not sure how many of them remembered my statement on hearing the President of the United States declare that Poland (and presumably other satellite countries) was not under Soviet domination. A rather strange evaluation of freedom which, as events were to show, greatly benefited Carter, according to his own admission. The twentieth anniversary of the Hungarian Revolution was only two weeks away and less than a year away was President Carter's decision to return to Hungary Saint Stephen's crown, the thousand-year-old symbol of that nation's independence!

58. Their statements to that effect are quoted in my Gifford Lectures, *The Road*

of Science and the Ways to God (Chicago: University of Chicago Press, 1978), with due emphasis on the point that endorsement of the idea of a strictly coherent universe has always been the hallmark of theistic thought, or of ways of thinking germane to it.

59.   In speaking of the high degree of cosmic order which we cannot expect a priori, Einstein remarked in his letter of March 30, 1952, to his friend, M. Solovine: "And here is the weak point of positivists and of professional atheists, who feel happy because they think that they have not only pre-empted the world of the divine, but also of the miraculous. Curiously, we have to be resigned to recognizing the 'miracle' without having any legitimate way of getting any further. I have to add the last point explicitly, lest you think that, weakened by age, I have fallen into the hands of priests." A. Einstein, *Lettres à Maurice Solovine*, reproduits en facsimilé et traduites en français (Paris: Gauthier-Villars, 1956), p. 115. In his letter of Jan. 1, 1951, also to Solovine, Einstein wrote: "I have never found a better expression than the expression 'religious' for this trust in the rational nature of reality and of its peculiar accessibility to the human mind. Where this trust is lacking science degenerates into an uninspired procedure. Let the devil care if the priests make capital out of this. There is no remedy for that." (ibid., pp. 102–103). The trust in the rationality of the cosmos which Einstein had in mind was not, of course, the facile trust of a priori thinkers in their own minds. Although he had repeatedly and emphatically endorsed the freely created character of concepts most useful in science, he never lost sight of the need of their empirical verification. He thus put himself in his nonphilosophical way somewhere in the epistemological middle between idealism and empiricism, a fact full of far-reaching metaphysical implications, and a point which Mach was the first to recognize to the extent of parting ways with the architect of relativity, special and general.

60.   See A. Heidel, *The Babylonian Genesis: The Story of Creation* (2nd ed.; Chicago: University of Chicago Press, 1951), p. 101.

61.   For a discussion of this point in its relation to the future fortunes of the scientific enterprise, see my *Science and Creation: From Eternal Cycles to an Oscillating Universe* (Edinburgh: Scottish Academic Press, 1974), Chapters IV and VI.

62.   As emphasized in the section of my Gifford Lectures mentioned in note 12 above.

63.   "On the Reconcentration of the Mechanical Energy of the Universe" (1852), in *Miscellaneous Scientific Papers*, edited by W. J. Millar (London: Charles Griffin and Co., 1881), pp. 200–202.

64.   The best modern example of these "bootstrap" cosmologies is the steady-state theory in which the *creation* of new matter out of nothing is credited to the expansion of the matter constituting the observable universe. It made the philosopher of science, M. Bunge, muse in 1962 without sus-

pecting how timely his musing would be as late as 1976: "The steady-state theory was philosophically killed several years ago by M. K. Munitz, *Brit. Jour. Phil. Sci.*, 5, 32 (1954), but apparently this criticism has been as ineffective as Augustine's refutation of astrology in his *Confessions*" ["Cosmology and Magic," in *The Monist*, **47** (1962), p. 126, note].

The Place of Facts
in a World of Values

by

HILARY WHITEHALL PUTNAM
Beverly Pearson Professor of Modern Mathematics
and Mathematical Logic
Harvard University

Science tells us—or at least we are *told* that "science tells us"—that we live in a world of swarming particles, spiraling DNA molecules, machines that compute, and such esoteric objects as black holes and neutron stars. In such a world, where can we hope for meaning or for a foundation for our values? Jacques Ellul[1] tells us—and I think he is right—that the themes of the present day are science and sexuality. He also tells us that in modern secular society most people—the people who think of themselves as "enlightened," in fact—are caught in a peculiar contradiction. On the one hand, nothing is regarded as more *irrational* than Christianity (or Judaism). On the other hand,

Europeans and Americans are going in droves for every kind of pseudoreligion one can think of. And the *verboten* desire for religion (*verboten* as "irrational," "unscientific"—the two words are treated as synonyms) does not only break out in the form of a million and one new and revived cults; it breaks out even more alarmingly in the form of a certain religionizing of the political—even "middle of the roaders" rarely discuss political questions any more without a special kind of commitment, more appropriate to the defense of a faith than to the discussion of public policy. At the same time, Leszek Kolakowski[2] writes despairingly (only he would not agree that this is despairing) that "the gulf between normative and empirical knowledge" cannot be bridged, and "the former can be justified only by the force of tradition and myth." But if Kolakowski is right, and at the same time "tradition and myth" are in vast disrepute, what then? Dostoevsky's "if God is dead then everything is permitted" may have been invalid logic, but accurate sociology. Indeed, looking at the world in which we live—this Babylon!—who can doubt it?

But a word of caution is perhaps not out of place. *Many* world views—even if some of them denied that they *were* world views—have been advanced in the name of science in the past two centuries. At one time Evolutionism, by which I mean not the theory of evolution but the doctrine that the fate of Man is to evolve and evolve until Man is virtually God—a doctrine which appeared on the scene at least 50 years *before* the theory of evolution, and which permeates the thought of Marx and Spencer (and perhaps even so moderate a liberal as John Dewey?)—was supposed to be *the* scientific view of things. More recently, I have heard extreme pessimism advanced for the same role—John von Neuman told me before he died that it was absolutely certain that (1) there would be a nuclear war; and (2) everyone would perish in it (let us hope that he was not right about both predictions). To move from such moral and human questions to more abstract metaphysical questions; both idealism (rebaptized "phenomenalism" or "positivism"), that is, the view that all there really is (or all that can be spoken of, anyway) is *sensations* and similar mental phenomena, and materialism (rebaptized "scientific realism"), have been enthusiastically espoused as *the* philosophy of Science. And both the view that there is an unchanging *scientific method*, and the view that what science *is* is itself all historical and relative have been held with the same enthusiasm.

Moreover, the metaphysical view that laymen widely assume to be the view dictated by modern science—the view that it is all atoms swarming in the void, and that there are no objective values, only swarming instincts and desires and the interests of various groups—does not even have the virtue of novelty. Lucretius already thought it was all atoms swarming in the void, and ethical relativism and scepticism were well known to the Greeks of Plato's day. If science really does reveal to us what our metaphysics and moral outlook should be, its revelation is neither monolithic nor new. Some revelations in history have had undeniable dignity and beauty, even for those of us who disbelieve in them; is it not rather sad, even a bit Prufrockian, if the *final* revelation turns out to consist of a half a dozen ideas which are not mutually consistent (you have to take your pick), and each of which looks, well, just a little bit *half-baked*?

In my talk today, I want to defend the view that "scientific" is *not* coextensive with "rational." There are many perfectly rational beliefs that cannot be tested "scientifically." But more than that, I want to defend the view that there are whole domains of fact with respect to which *present day* science tells us nothing at all, not even that the facts in question exist. These domains are not new or strange. Three of them are: (1) the domain of objective values; (2) the domain of freedom; (3) the domain of rationality itself.

I want to say something about each of these domains in turn.

OBJECTIVE VALUES

Let me begin on a personal note. My training as a philosopher of science came from logical positivists (or "Logical Empiricists," with a capital "L" and a capital "E," as they preferred to style themselves). These men firmly believed in the "emotive theory of ethical discourse"—that is, they firmly believed that the sole function of ethical discourse is to express feelings or acts of will or, more vaguely, "attitudes."

Although it is not *quite* right to say that these men thought values were matters of "subjective taste," which is how any layman would describe this view to any other layman—and, indeed, how else *can* it be described in language that *tout le monde* can understand?—it is essentially right. Choosing a morality is choosing a way of life, Professor Hare of Oxford tells us. While I rather think he uses "choice of a way of life" as a mystifying phrase, his too is a correct expression of this view, provided we think of choosing a way of life as something like choosing, say, an outdoor life as opposed to a life of listening to classical music and engaging only in sedentary pursuits (note that a "way of life" in the *literal* sense is something that can be unalterable for a person, something he can be passionate about, and something he can be—though he need not be—intolerant about).

For several years after I got my doctorate I firmly believed this view or a sophisticated variant (at least I thought it was sophisticated) of my own devising. I thought that something was good in the specifically moral sense if it "answers to the interests associated with the institution of morality"—a view suggested by the analysis of the meaning of the word *good* in Paul Ziff's book *Semantic Analysis*, though not specifically contained in that book. The advantage of this view over the older emotivism was that it allowed me to say that value judgments were *true* or *false*—to find out if a judgment of the form X *is good* is true one just has to discover if X answers to "the interests associated with the institution of morality." *How* one discovers what those interests are, I was never able to answer to my own satisfaction. But, this difficulty aside, I never doubted Professor Hare's view just alluded to—although I disagreed with his semantic analysis of *good*—the view that the decision to try to be or do good is just a "choice of a way of life," namely, to subscribe or not to subscribe to an "institution."

Anyway, in the middle of this period I found myself with a severe moral problem in my own life (what it was I am not going to tell you, nor would it be particularly relevant as far as I can see). And the interesting thing is that I found myself agonized over whether what I was doing, or contemplating doing, or had done, was *right—really* right. And I did not just mean whether it was in accord with the Utilitarian maxim to do what will lead to the greatest happiness of the greatest number (although I thought about that), but whether, if it was, then was that the *right* maxim for such a case? And I do not think I meant would some semantic analysis of the word "good," or some analysis of "the institution of morality," support what I was doing. But the *most* interesting thing is that it never occurred to me that there was *any* inconsistency between my meta-ethical view that it was all just choice of a "way of life" and my agonized belief that what I was doing had to be either *right* or *wrong*. (I do not mean that there are no borderline cases; I felt that in *this* case what I was doing was either right or wrong.)

I would not have you think that this inconsistency was peculiar to me, however. My emotivist teachers and colleagues—in particular Hans Reichenbach and Rudolf Carnap—were fine and principled human beings. Both had been anti-Nazi when, to the shame of the philosophical profession, some world famous German philosophers succumbed to Nazi ideology, and both were generous and idealistic men, wonderful to their students. I am sure that both had deep convictions about right and wrong—convictions that they would have laid down their lives rather than betray.

But the most charming example of this sort of inconsistency occurs in a famous lecture by Frank Ramsey, the British philosopher of the 1930s who was regarded as so brilliant by such men as Russell and Wittgenstein, and who died while still not 30 years old. The lecture[3] wittily defends the thesis that "there is nothing to discuss"—that is, there is nothing that can sensibly be discussed any more *by laymen*. As Ramsey put it, "Let us review the possible subjects of discussion. They fall, as far as I can see, under the heads of science, philosophy, history and politics, psychology, and aesthetics; where, not to beg any question, I am separating psychology from other sciences.

"Science, history, and politics are not suited for discussion except by experts. Others are simply in the position of requiring more information; and, till they have acquired all available information, cannot do anything but accept on authority the opinions of those better qualified. Then there is philosophy; this, too, has become too technical for the layman. Besides this disadvantage, the conclusion of the greatest modern philosopher (*Ramsey is referring to Wittgenstein*—H.P.) is that there is no such subject as philosophy; that it is an activity, not a doctrine; and that, instead of answering questions, it aims merely at curing headaches. It might be thought that, apart from this technical philosophy whose centre is logic, there was a sort of popular philosophy which dealt with such subjects as the relation of man to nature, and the meaning of morality. But any attempt to treat such topics seriously reduces them to questions either of science or of technical philosophy, or results more immediately in perceiving them to be non-sensical."

Ramsey takes as an example a lecture of Russell's. He remarks that Russell's philosophy of nature "consisted mainly of the conclusions of modern physics, physiology, and astronomy, with a slight admixture of his own theory of material objects as a particular kind of logical construction." This, he points out, is something that can only be discussed "by someone with adequate knowledge of relativity, atomic theory, physiology, and mathematical logic." Then he adds, "His philosophy of value consisted in saying that the only questions about value were what men desired and how their desires might be satisfied, and then he went on to answer these questions. Thus the whole subject became part of psychology, and its discussion would be a psychological one.

"Of course his main statement about value might be disputed, but most of us would agree that the objectivity of ethics was a thing we had settled and dismissed with the existence of God. *Theology and Absolute Ethics are two famous subjects which we have realized to have no real objects.*" (Italics mine.)

This is, for once, an unvarnished and unsoftened statement of the view that there are *no* objective ethical judgments. But let me quote you the concluding words of Ramsey's charming lecture:

"I find, just now at least, the world a pleasant and exciting place. You may find it depressing; I am sorry for you, and you despise me. But I have reason and you have none; you would only have a reason for despising me if your feeling corresponded to the fact in a way mine didn't. But neither can correspond to the fact. The fact is not in itself good or bad; it is just that it thrills me but depresses you. On the other hand, I pity you with reason, because it is pleasanter to be thrilled than to be depressed, and *not merely pleasanter but better for all one's activities.*" (Italics mine.)

So even Ramsey finds *one* judgment of value to be a judgment of *reason*!

But is the emotivist really inconsistent? True, the man on the street naturally assumes that, if someone really thinks all moralities are just expressions of subjective preference, choices of a "way of life," then that person must agree that no morality is better or worse than any other, that all moralities are equally legitimate. But that does not strictly follow. The emotivist can reply that words like "legitimate" are emotive words too! The emotivist can say that torture is wrong (or, on the contrary, that it is right when the nation needs to find out the military secrets of an enemy, or whatever) and *also* say that any other point of view about the matter is *bad, evil, monstrous, not legitimate*—anything except that the point of view he opposes is *false*. (Why he does not say that *true* and *false* are emotive too is not quite clear.)

Yehuda Elkana[4] has spoken of "two-level thinking" on the part of some scientists, philosophers, and so on. The "two-level thinker" speaks like a realist on the "first-order" level (when he is talking about things other than his own talk). For example, he says "electrons are flowing through this wire," or "there is an inkwell on the table," or "torture is wrong." But he denies the objectivity of his own first-level talk (i.e., that it possesses truth or falsity, except relative to a culture, or whatever) when he comes to comment philosophically on his own first-level talk. On the "meta" level he says things like "electrons are just permanent possibilities of sensation," or *"right* and *wrong* are just performatory locutions for performing the speech acts of prescribing and proscribing." In Elkana's terminology, the emotivist can avoid the charge of inconsistency by saying that he is engaging in two-level thinking in ethics.

While it is true that the emotivist can avoid the charge of strict formal inconsistency in this way, it seems to me that the layman's feeling that something "inconsistent" is going on possesses a deep basis.

For one thing, on such a conception, what sense does it make to *worry* if something is right or wrong, as I did? Of course, one can worry about the facts (whether something really conduces to the greatest happiness of the greatest number, or whatever). One can worry about whether an action really agrees with a maxim; but how can one worry about whether or not one has the right maxim? It would seem that one should say to oneself in such a case "you must simply *choose* a morality." The idea that there is such a thing as being right or wrong about which morality one chooses must just be a hangover from absolute ethics—that "subject which we have realized to have no real object."

And what could our *motive* for being ethical possibly be? The favorite categories of motivation employed by *scientistic* thought are *instinct* and *conditioning*. I see no evidence whatsoever that being ethical is an instinct. My friend E. O. Wilson has recently suggested that ethics may be based on altruistic and gregarious instincts which are themselves the product of natural selection. (He is very careful not to derive a theory of what we should do from this.) However, whatever may be the *origin* of our altruistic and gregarious impulses, our loyalty, and so on, it is not the case that following these impulses is the way to be ethical. There is no instinct or instinctual impulse which may not lead to great evil if followed to extremes, as moralists have long known. Altruism, insofar as it is just an instinct or instinctual impulse, may lead one to kill the old people in the tribe for the sake of posterity (more food for the young); or to torture someone to discover the military secrets of an enemy state; and so forth. Even feelings of kindness may lead one to do wrong, for example, if one's sympathy for a hunted person leads one to allow a Martin Bormann to escape deserved punishment, or one's sympathy for the workers leads one to set wages so high that disastrous inflation results. And there is no instinct which it may not be one's duty to follow in some situations. The whole idea that our natural instincts can be classified into good and evil instincts, or our natural emotions into good and evil emotions is an error, as moral writers have again pointed out.

But suppose that the motivation for ethical behavior *is* an instinct. (Imagine that sociobiology discovers the instinct.) Why should we not *suppress* it? We suppress all of our other instincts sometimes; indeed, we have to to avoid disaster. Why should we not suppress our instinct to be ethical, whenever following it will lead to loss of our life, or loss of a loved one, or other hardship to ourselves? Why should we not be moral only when it does not cost too much, or when it does not pay too well to be immoral?

Nor does appeal to "conditioning" help matters any. What this comes to is either (1) the claim that we act ethically because we cannot help it (our toilet training, etc., just left us with this "super ego structure," according to the Freudian version— Freud, in particular, is often credited with a "Copernican Revolution" in psychology for giving just this explanation of our conscience), or (2) the claim that we act ethically because we want our neighbors to approve of us. [Of course, (1) and (2) are not incompatible, and both (1) and (2) assume instincts as well as conditioning.] But it is just false that we cannot help acting ethically! We do have a choice—and it is often terribly hard to act ethically. The fact of both personal life and human history is that people do not act ethically, not that they do. And the desire for the approval of our neighbors can be worthy or unworthy. Many people do wrong things because they want the approval of their neighbors. If wanting to be ethical is like wanting an expensive and vulgar house because the neighbors will be impressed, why should anyone bother?

Let me try to make the problem vivid by means of an example. (I lack the literary skill required to do it properly; if you want to see this sort of example really developed in full human richness I would urge you to read Kamala Markandaya's moving novel *A Handful of Rice*.) Imagine a poor peasant boy growing up in poverty-stricken Sicily (Markandaya's novel is set in even more poverty-stricken Calcutta). Let us suppose he is offered the opportunity to become a member of the Mafia. If he accepts, he will do evil things—sell terrible drugs, run prostitution and gambling rackets, and even commit murders; but he will also live comfortably, have friends and women, and, perhaps, even enjoy a kind of respect and admiration. If he refuses, he will live a life of almost unimaginable poverty, will probably see some of the children he will have die for lack of food and medicine, and so on.

I am not supposing that the boy will be perfectly happy as a *mafioso*. Perfect happiness is not one of the options here. Notice, please, that this kind of sacrifice—and it is just as real as and even more bitter than the sacrifice on the battlefield, or the public execution for one's convictions—is one that millions of people, millions of the poor, make and always have made. And unlike the sacrifice on the battlefield or the public execution, there are no posthumous medals, no stories told to the children and grandchildren—some of the bitterest sacrifices that people make for what is right are (and, indeed, must be) taken as simply a matter of course by their neighbors and friends.

Now I ask you, would anyone make such a sacrifice if he believed that the thing that was impelling him to make it was, at bottom, just a desire to impress (some of) his neighbors, or even in the same ballpark as a desire to impress the neighbors? or his toilet training? It is all very fine for comfortable Oxford professors and comfortable French existentialists to wax rhetorical about how one has to "choose a way of life" and commit oneself to it (even if the commitment is "absurd," the existentialists will add). And this rhetoric really impresses people like ourselves, who are reasonably prosperous. But the poor person who makes such a sacrifice makes it precisely because he does not see it that way. Would any one *really* choose such a life if he thought that *all* it was was "a choice of a way of life"? Of course, he makes the choice he does because he knows that that choice is his *duty*. And he knows that he cannot choose his duties, at least not in this respect.

The situation is this: the popular line of thought today has no room for the traditional notion of reason as a faculty that dictates *ends* to us and not only means to ends dictated by instinct and modified by conditioning. Hume's dictum that "reason is and ought to be the slave of the passions" expresses the modern idea exactly—reason *equals* instrumental reason. But if reason *cannot* dictate ends, then there cannot be a reason why I should go *against* my "passions," except the inadequate reason of another "passion." There cannot be such a thing as a *rational* answer to "why should I do my duty?" And an irrational answer which is admitted and recognized to be irrational or arational is no good.

The fact is that the knowledge that there are objective values involves not merely the knowledge that moral judgments are true or false, but just as importantly the knowledge that what regulates the behavior of the person who acts from the motive of duty, what directs him to suppress this instinct now and grafity it then, to withstand and resist this kind of conditioning when his neighbors attempt it, but not to object to this other kind, is neither mere "instinct" nor mere "conditioning."

Recently Iris Murdoch[5] has put forward the view that the whole "fact/value" dichotomy stems from a faulty moral psychology: from the metaphysical picture of "the neutral facts" (apprehended by a totally *uncaring* faculty of reason) and the will which, having learned the neutral facts, must "choose values" either *arbitrarily* (the existentialist picture) or on the basis of "instinct." I think Miss Murdoch is right; but setting this faulty moral psychology right will involve deep philosophical work involving the notions of "reason" and "fact" (as she, of course, recognizes).

FREEDOM

The problem of the freedom of the will has many aspects, as I am sure many of you realize. I am going to talk about only one aspect: our belief that we could have done otherwise in certain situations, for example:

> *I could have refrained from boasting on that occasion,*
> *I could have taken a different job,*
> *I could have spent my vacation in a different place.*

One of the first English speaking philosophers to try to reconcile freedom and determinism was David Hume. Although in the *Treatise* Hume denies the freedom of the will, in the *Inquiry* he puts forward the theory that the incompatibility is only apparent. One of the simplest statements of this sort of view (like all philosophical positions, it has versions which are almost infinitely complicated) is the following: "we could have acted otherwise" is false when the causal chain that ended with our action (or with the trajectory of our body that was the "physical realization" of our action) was of one kind—when it did not "pass through our will," in the simplest version of the theory—and is true when the causal chain was of another kind—when it did "pass through the will." If someone ties me up and carries me to Vienna, then it is false that I could have done otherwise than go to Vienna; but if I decide to go to Vienna, then buy the ticket, then take the plane, and so on, and the causal chain is of the usual type for a "voluntary" action, then I could have done otherwise. (The Hegel-Marx dictum that "Freedom is obedience to Law" is a variant of the same idea, I believe.)

There is one obvious difficulty: asking whether it is true that an event could possibly have failed to happen is just not the same question as whether its cause was of kind X or kind Y! Hume is just substituting a different question for the question that was asked.

It is true that in ordinary language we often use "can" and "could" in senses that do not imply full possibility. Sometimes "John can do X" just means John has the ability (nothing said about opportunity); sometimes it means "John has the opportunity" (nothing said about ability). But that is not at issue. What we are concerned with is whether we "really actually fully could" have done otherwise, in John Austin's phrase.

Consider the case of a man, call him McX, who suffers from a debilitating compulsion neurosis. McX cannot bear not to be in contact with a wall; walking across an open courtyard without touching any wall is not possible for him (a famous English philosopher had such a compulsion). Is it true that McX "has the ability" to walk across the courtyard without touching any wall?

In one sense it is: McX is not paralyzed and has learned to walk. In another sense it is not: McX breaks into a cold sweat at the mere thought, no matter how hard he tries he "just can not do it," and so on. "Ability" talk is not helpful here; what is clear is that McX *cannot* walk across the courtyard without touching a wall.

Similarly, if *events out of my control* determine with physical necessity that I will pick up a newspaper at time *t*, then it is false that I could have refrained from picking up the newspaper at time *t*. "I could have done something else" in the sense of having the ability, in one sense of "having the ability," just as McX was "able" to cross in the sense of having learned to walk; "I could have done something else" in the sense of having had the opportunity, just as McX "could have crossed" in the sense of having had the opportunity; but in the sense of "really actually fully could have," it is just *false* that I could have done something else, on the supposition that determinism is true.

This simple and correct argument has been challenged however. Some say that the argument is fallacious; that it exhibits the invalid form

Necessarily (if p *then* q)

p

_____

(Therefore) Necessarily q.

Is this the case?

It is not. The above argument form is clearly fallacious when "necessarily" is taken in the sense of "it is a necessary truth that" (either in the sense of logically necessary or in the sense of physically necessary). Thus it would be fallacious to argue:

It is a necessary truth that all triangles are three sided
This table is a triangle (in shape)

(Therefore) it is a necessary truth that this table is three sided.

But the argument that:

Necessarily (If the state of the world at time t *was S, then I will*
   pick up the newspaper)
The state of the world at time t *was S*

(Therefore) It is impossible *that I will not pick up the newspaper*

is not of this form. Saul Kripke[6] has given an elegant rejoinder to the claim that the arguments are of the same form (and therefore the second argument, like the first, is fallacious). Kripke simply pointed out that if that is all there is to the second argument, then the following everyday argument must *also* be fallacious:

Necessarily (If I miss all the trains to London then I cannot get to
   London today)
I have missed all the trains to London

(Therefore) It is impossible for me to get to London today

—but this argument is clearly correct!

Moreover, Kripke points out, it is easy to give a modal semantics that justifies this pattern of argument (Kripke is the greatest modern authority on modal logic). Consider a world in which there are branching futures. Call the tensed statement "It is possible that X will occur" true as long as there is a branch in the future that leads to the occurrence of X, and false as soon as the last branch leading to the occurrence of X has been passed. It is easy to check that with this semantics—which is just the intuitive semantics for tensed possibility statements—both the "last train to London" argument and the argument that determinism is incompatible with freedom are perfectly correct.

However bad this incompatibility might have been in the heyday of Newtonian physics, the important fact is that here it was the science that turned out to be wrong and not the belief that we are free in the sense that we could have done otherwise. Present day physics is indeterministic, not deterministic, and there is absolutely no incompatibility between indeterministic physics and the sort of statement we listed. I am not claiming that "could" applied to a human action *just* means "compatible with physical law and antecedent conditions"; indeed, I am not proposing an analysis of "could" at all. I am saying that the *case* for incompatibility has dissolved; further problems of philosophical analysis will, of course, continue to be with us.

Now, one might expect that philosophers would have hailed this result. Should we not be glad if something we seem to know so clearly as the fact that we could have spent our vacation at a different place than we did has turned out not to be incompatible with well-confirmed scientific theory? But, in fact, with very few exceptions, philosophers have scoffed at the significance of quantum mechanical indeterminacy for the free will problem! The reason, I think, is that the philosophers find anything but mechanism an embarassment; they are in the grip of a fashion, and the only thing more powerful than reason in the history of philosophy is intellectual fashion. One exception to this regrettable tendency is Professor Elizabeth Anscombe: in her powerful inaugural address[7] she both recognizes the importance of the determinism issue and the importance of the fact that the scientific evidence does not any longer support determinism, if it ever did.

One way of scoffing at the significance of indeterminism is to pretend that it makes no difference to "ordinary macroscopic events" such as the motions of human bodies. This is an outright mistake, and Professor Anscombe disposes of it with great elegance. Another way is to say that we are "no better off" (in terms of moral responsibility for our actions) if our actions are "the product of chance" than if they are determined.

This claim changes the question, of course. We were not talking about moral responsibility but about freedom; and while these are related (no freedom, no moral responsibility, at least in Kant's view) they are not the same. The original problem was an incompatibility between deterministic physics and freedom; and we have seen that that does not exist any longer. I know of no argument whatsoever that there is an incompatibility between *indeterministic* physics and freedom. But responsibility is important too. What reason is there to think that there is an incompatibility between indeterministic physics and responsibility?

The argument is that indeterministic physics says that our actions (or at least their component bodily motions) are produced by chance. And how can we think of ourselves as a kind of roulette wheel and still ascribe moral predicates? But it is just false that indeterministic physics says that our actions are "produced by chance"—that is, *caused* by Chance (with a capital "C").

Let us be clear on this. Indeterministic physics uses the notion of *probability*. And while the analysis of the notion is still in dispute, there is no reason to interpret probability as Aristotelian Chance (which *was* a cause—a cause of whatever is unexplainable).

If we stick to what is generally agreed on among scientists, all we can say "probability" means is *the presence of statistical regularities*. Now, no philosopher ever doubted that there are statistical regularities in human behavior, even if the terminology "statistical regularity" was not always available. No one ever doubted that, for example, there are true statements of the form *90% of people with such-and-such a temperament and upbringing tend to succumb to such-and-such a temptation.* This statement does *not* say that each individual succumbing is "produced" by something called "chance." It does not say anything about the individual event *at all*.

Nor does it help to say our actions are "random" rather than "chance." There are many concepts of randomness used in statistics; they all have to do with the distribution of frequencies in subsequences of the main sequence. None of them means "produced by chance." The incompatibility is *not* between indeterministic physics and moral responsibility; it is between moral responsibility and a metaphysics—the metaphysics of Chance.

The suggestion that quantum mechanical indeterminacy may, after all, bear on questions of responsibility is supported by the following reflection. Assume determinism is true, and assume—what is extremely plausible—that very few people who commit evil deeds are determined to commit them by their genes *alone*. (Indeed, if someone had *such* bad heredity that he was bound to commit crimes no matter how he was brought up or in what circumstances he found himself, we would be inclined to *excuse* him on just that account.) It is part of our contemporary moral sensibility that certain environmental conditions are (at least partly) *excusing*. For example, someone raised in poverty, especially if raised by cruel and unloving parents, deprived of proper moral guidance, and so on, seems to us less blameworthy than a rich, "advantaged" person who does wrong. Yet, if our supposition is correct, the rich, "advantaged" person is also the "victim" of a pattern of environmental circumstances—one that we do not have a *name* for (like "poverty"), and, indeed, one we do not recognize. It is also true of the rich, "advantaged" person that if he or she had had a different upbringing and environment he or she would not have become evil. So, in this sense, determinism does threaten our entire way of thinking about moral responsibility and excusability.

On the indeterminist picture, the situation is different. What similarity is there between the "advantaged" person who does something he might well not have done and the disadvantaged person who does something he had virtually no chance of not doing (or performs an act such that he was virtually bound to perform some act of that *type*, if not that token action itself)? The answer "they were both caused to do what they did by Chance" is a kind of metaphysical joke. There is a difference between real environmental conditions and metaphysical bogeys.

REASON

I want to say something now about a very old problem in philosophy—Hume's problem of induction. One way of stating the problem is this: we believe that the future will resemble the past—not, of course, in every respect, but in the respect that the statements we call "laws of nature" will continue to be true in the future (or *as* true as they have been in the past). Some scientists think that even this may not be true; that in billions of years the "laws of nature" may change, or at least the values of the basic physical constants they contain may change; but I shall not be concerned with such a vast time scale as billions of years. If we confine ourselves to the next hundred, or thousand, or even million years, then we do believe that the laws of nature will be the same—at least the basic laws of physics and chemistry (and it is not clear that the other sciences, such as sociology, *have* "laws" in the same sense as physics and chemistry do). Also, we believe that the universe will continue to consist of electrons, neutrons, protons, and so on. There are other very general respects in which we assume the future will resemble the past; but for our purposes these two will suffice. If this is false, if at some future time *t* the universe stops obeying the laws it has always obeyed, stops even consisting of the same elementary constituents it has always consisted of, we may speak of a global catastrophe. The assumption that the future will resemble the past means that there will not be a global catastrophe (in the next thousand, or whatever, years). How do we know that this is true or even probably true?

Some people say that if there is going to be such a global catastrophe, we cannot do anything about it anyway, so we are justified in ignoring the possibility; but it is not true that "we cannot do anything about it anyway" (unless we assume the global catastrophe would wipe out life, which is not necessarily the case—remember, the *laws* are going to be different!). Suppose that in California there is an occultist sect which in fact predicts a global catastrophe for next year, and which pretends to tell us exactly what we must do to survive it. Then there is something we can "do about it," if that sect is right—we can all join that sect and follow its instructions. It is not as if the human race could not help being scientific—that is hardly the case!—or was incapable of joining and following cults. What is really meant by the statement that "we cannot do anything about it" if there is going to be a global catastrophe is that we cannot have *good reason* to believe such a cult rather than any other cult or fad or school of thought. But do we have *good reason* to believe what we do believe? Is there such a thing as good reason?

Some philosophers nowadays say that reasonableness is just "coherence"; our beliefs are reasonable just in the case that our experiences, methods, and beliefs all mutually "cohere" or support and accord with one another. I do not know just what "coherence" is (nor do I know where the *criteria* of "coherence" are supposed to come from—do they too only have to "cohere"? If so, anyone can reasonably believe *anything*, provided he just has the right notion of "coherence."). But, assuming that we are allowed to use our natural judgments of what is plausible and implausible in judging "coherence" (and one cannot reason at all if one tries to stand outside of every tradition of reasoning), let us ask whether our beliefs *are* all that "coherent" with respect to the future resembling the past.

In one sense they are. We believe a great many *specific* laws will continue to be true (or as true as they are now) in the future; and each of these beliefs supports the general belief that there will not be a global catastrophe. But each of these specific beliefs was confirmed *assuming* there would not be a global catastrophe. This only shows that our belief that the future will resemble the past is a very fundamental belief—so fundamental that all specific beliefs are adjusted to it. In itself this is good, not bad—we would expect a coherent system to have some fundamental assumptions, and some kind of logical structure. But one more question: what is the *cause* of our belief that the future will resemble the past? Knowledge is part of the subject matter of our knowledge; we should expect a really coherent system to include some plausible account of how we know that system itself to be true, or approximately true.

The question is, of course, very strange. Traditionally, philosophers thought beliefs had *reasons,* not *causes*. But it is an essential part of the story that belief is itself a formation of brain traces, or something of that kind, and that it is *caused*—caused by natural selection plus cultural conditioning. We are caused to believe that the future will resemble the past—caused, no doubt, by the fact that that belief has had survival value *in the past*.

But that means our belief that the future will resemble the past is exactly like the belief (or rather expectation) that dinosaurs doubtless had that conditions would continue to be excellent for dinosaurs! There is *no* general law that what had survival value in the past will continue to have survival value in the future! In the words of the philosopher George Santayana, our belief that the future will resemble the past, and indeed all of our "inductive" beliefs, are based on *animal faith*.

But belief that the future will resemble the past is at least *reasonable* (even if we seem to have trouble in saying why). I want to talk now about some other things that we believe that are not even reasonable.

By all the evidence, homo sapiens, essentially as we know him today, was already on the planet 30,000 years ago. At that time there was no civilization, no mathematics beyond, perhaps, a little counting, and no technology beyond stone axes and fires. Yet our intellectual abilities were, according to evolutionists, essentially as they are today—a cro-magnon or homo sapiens boy or girl of that period, if transported to the present day, could learn calculus or physics, perhaps even become a scientist or inventor. According to the current scientific view, all this latent intelligence was the result of natural selection.

Think about what this means. We are, on this view, computing machines programmed by blind evolution—computing machines programmed by a fool! For evolution *is* a fool as far as knowing about calculus, or proving theorems, or making up physical theories, or inventing telephones is concerned. A fool, but not a moron—evolution is very smart as far as the behavior of deer or manipulating stone axes is concerned. What I mean is this: selection pressure would naturally weed out members of the different humanoid species who could not learn to hunt deer, use stone axes and spears, make fires, or cooperate in a hunt. In that sense, being programmed by blind evolution (selection pressure) is like being programmed by someone who has a good idea of what it takes to make a fire, hunt deer, and so on. It is as if we had a maker who knew a lot about deer, fires, hunts, and so on. But there was no selection pressure for being able to make up scientific theories—that was not something *any* member of the several humanoid species did *then*. So that a device molded by selection pressure to do the one thing should be able to do the other is quite a miracle—exactly as if we set a fool (who knew about hunting with stone axes and spears, etc.) to program a computer and he programmed it so that it could *also* discover the theory of relativity.

Sometimes the suggestion is advanced that our abilities to do mathematics, discover physical laws of an abstract kind, and so forth are the result of concomitant variation and not selection pressure. (What this means is that sometimes an organ appears because the DNA instructions for manufacturing that organ happened to be carried on a gene that also carried something else that did have survival value. So the species acquired the organ; but the survival value of the organ was quite irrelevant.) After all, it is pointed out, we do not know what constraints nature operated under in packing billions of neural connections into a small brain pan.

Again, think about what this means. Nature—which imposed the constraints on how neural connections could be manufactured and installed—does not even have the pseudo-"intelligence" that selection pressure does; that is, nature has no *ends*, and does not simulate having any. (Selection pressure at least *simulates* having ends.) So it is as if selection pressure—the fool that programmed us—had to operate subject to constraints imposed by a *moron*. And it just happened that the moron imposed such lucky constraints that the only way the fool could program us involved making us intelligent enough to do mathematics, discover general relativity, and so on! The theory that we are computers programmed by a fool operating subject to the constraints imposed by a moron is worse, not better, than the "selection pressure" theory. And "concomitant variation" is just a fancy name for *coincidence*.

At this point I imagine that an objection will occur to you. Is it not our *intelligence* that accounts for the ability of our species to do natural philosophy, mathematics, and so on? And does not intelligence have survival value if anything does?

The answer is that intelligence, in the sense of the ability to use language, manipulate tools, and so on, is not enough to enable a species to do science. It also has to have the right set of *prejudices*. But let me explain.

Suppose we had evolved with all the intelligence that we have (whatever "intelligence" is), but with a firm prejudice against the unobservable. Suppose we only believed in things we could see, hear, feel, touch, and so forth. (We would make an excellent race of logical positivists.) We would not believe in gods, spirits in the trees and in the rivers, substance and accidents, forces and "natural motions," and so on. We would never develop a religion or a metaphysics. But as far as observable things are concerned, we could be as "scientific" as you please. We might even be more "rational" than humans are, because we would not be led astray by "metaphysical prejudices." Then we—or such a race of beings—could hunt deer, use stone axes and spears, make fires, and so forth, just as well as we actually can. Such a race might even develop a civilization to the level of ancient Egypt. It would not develop geometry, beyond the Egyptian level of practical land measurement, because the notion of a straight line with no thickness at all, or the notion of a point with no dimensions at all, would make no sense to it. It would never speculate about atoms swarming in the void, or about *vis viva*. And, interestingly enough, it would never develop physics or mathematics!

There are only two possibilities here. Either we postulate unobservable causes for observable events, and speculate about what these might be because it is *reasonable* to do so (but then what is reason?); or this is just a habit that we evolved with by *luck*. If, as positivists claim, *all* of our speculations about unobservables prior to modern science were *wrong*, as belief in objective values, God, and essences are supposed to be—or at best totally untestable prior to the nineteenth century, as belief in atoms swarming in the void was—then our incurable habit of speculating about unobservable causes for observable phenomena either had no survival value, or only had psychological value (it gave us comfort, and so on) which is irrelevant to and has no bearing upon truth and falsity. So then it is sheer *luck* that this metaphysical prejudice of ours, which we need to have to do science, evolved at all, or that we did not evolve with metaphysical prejudices which would prevent us from ever thinking of good scientific theories!

Although the point of view I am advancing here—that one cannot discover laws of nature unless one brings to nature a set of a priori prejudices which is not hopelessly wrong (one mathematical model for this idea is the idea of the scientist as "internalizing" a so-called "subjective probability metric" which assigns an antecedent probability—a probability *prior to experiment*—to hypotheses)—is becoming commonplace among inductive logicians, it is obvious why it should make some people feel uncomfortable. So an alternative account of the success of science is often proposed. This alternative account uses the notions of *trial and error* and *simplicity*.

After all, it is said, it took thousands of years before the theory of relativity was thought of. Maybe what happens is that we just test one scientific hypothesis after another until by trial and error we come to the right one. (Often it is added that we *must* eventually come to the true laws of nature, provided only that they are sufficiently "simple." This is where "simplicity" comes in.)

Now, the notion of simplicity is not really a very clear one. Almost *any* theory can be made to *look* "simple" if we are allowed to invent a notation specially for the purpose. And this is exactly what we do in science—we make up our notations so that our theories will look simple and elegant. It is not as if we were given a notation in advance and told "the laws of nature are expressible by simple and elegant formulas in *this* notation."

And, indeed, the history of science does not support the view that it was all trial and error, either in the sense of random trial and error or systematic search through all possibilities. Galileo discovered the Law of Inertia by thinking about and modifying fourteenth-century ideas, which themselves were a modification of Aristotle's ideas. It was a metaphysical line of thought that provided the general ballpark in which to look for a law of inertia. Einstein was working in the general ballpark provided by philosophical speculations about the relativity of motion, themselves centuries older than the evidence, when he produced the special theory of relativity. The general theory of relativity was suggested by geometrical analogies. And so on. There does not seem to be anything common to all the good theories that scientists succeeded in producing except this: each was suggested by some line of thinking that seemed *reasonable*, at least to the scientist who came up with the theory. (Of course, there is a great deal of disagreement about what seems reasonable, especially where "far out" ideas are concerned.)

I am suggesting that the old term "natural philosophy" *was* a good name for what we now call "science." Science is arrived at by reason, and not infallible, a priori reason that makes no mistakes, to be sure, but plausible reasoning that is often subjective, often controversial, but that, nevertheless, comes up with truths and approximate truths far more often than any trial and error procedure could be expected to do.

My purpose is not to argue that the theory of evolution is wrong. Nor am I covertly arguing for the doctrine of special creation. My own view is that the success of science cannot be anything but a puzzle as long as we view concepts and objects as radically independent; that is, as long as we think of "the world" as an entity that has a fixed nature, determined once and for all, independently of our framework of concepts.[8] Discussing this idea goes far beyond the bounds of *this* paper. But if we do shift our way of thinking to the extent of regarding "the world" as partly constituted by the representing mind, then many things in our popular philosophy (and even in technical philosophy) must be reexamined. To mention just two of them: (1) Locke held that the great metaphysical problem of realism, the problem of the relation of our concepts to their objects, would be solved by just natural scientific investigation, indefinitely continued. Kant held that Locke was wrong, and that this *philosophical* question was never going to be solved by empirical science. I am suggesting that on this subject Kant was right and Locke was wrong (which does not mean that science is unimportant in philosophy). (2) Since the birth of science thousands of years ago we have bifurcated the world into "reality"—what physical science describes—and appearance. And we have relegated aesthetic qualities, ethical qualities, psychological traits ("stubbornness," "patience," "rudeness"), and sometimes even dispositions and modalities to the junk heap of "appearance." I am suggesting that this is an error, and a subtle version of Locke's error. The "primary/secondary" or "reality/appearance" dichotomy is founded on and presupposes what Kant called "the transcendental illusion"—that empirical science describes (and *exhaustively* describes) a concept-independent, perspective-independent "reality."

The present talk has neither explained nor defended these views of mine. Its purpose has been more modest: to encourage those who hear it *not* to confuse science as it actually is—an ongoing activity whose results, spectacular as they are, are ever subject to modification, revision, and incorporation in a different theory or a different perspective—with any metaphysical picture that tries to wrap itself in the mantle of science; and to remind you that common sense and critical intelligence still have to be brought *to* scientific (as to all other) ideas; they are not a commodity to be purchased *from* "science." And, last but not least, I have tried to suggest that an adequate *philosophical* account of reason must not *explain away* ethical facts, but enable us to understand how they can be facts, and how we can know them.

NOTES AND REFERENCES

1. Cf. Jacques Ellul, *The New Demons* (New York: Seabury Press, 1975).
2. I quote here from an unpublished paper.
3. Cf. the final paper in Frank Ramsey, *The Foundations of Mathematics* (New York: Harcourt Brace, 1931).
4. Cf. Yehuda Elkana, "Science As a Cultural System," three lectures delivered to the Boston Colloquium in the Philosophy of Science in Fall 1976, forthcoming in *Boston Studies in the Philosophy of Science*.
5. Iris Murdoch, *The Sovereignty of "Good"* (New York: Schocken Books, 1975).
6. Kripke's remarks were made at *The Oxford International Symposium* held at Christ Church College, October 1976.
7. Reprinted in *Causation and Conditionals*, edited by Ernest Sosa, Oxford paperback, 1970.
8. Cf. Hilary Whitehall Putnam, "Realism and Reason," Presidential Address to the American Philosophical Association (Eastern Division), December 1976, reprinted in Hilary Whitehall Putnam, *Meaning and the Moral Sciences* (Routledge and Kegan Paul, 1977).